TRANSITIONAL MATHEMATICS
Developing Number Sense

STUDENT WORKBOOK

John Woodward, Ph.D. and Mary Stroh, B.S.

Sopris West Educational Services • Longmont, Colorado

Copyright 2004 by Sopris West Educational Services
All rights reserved.

No part of this work may be reproduced or transmitted in any form or by any means, electronic or mechanical, including photocopying or recording, or by any information retrieval system, without the express written permission of the publisher.

ISBN 1-57035-961-X

Edited by Beverly Rokes
Editorial assistance by Annette Reaves
Text design by Edward Horcharik *and* Sebastian Pallini
Text production by Edward Horcharik *and* Matthew Williams
Cover design by Sue Campbell
Production assistance by Denise Geddis

06 05 04 03   6 5 4 3 2 1

Printed in the United States of America

Published and Distributed by

SOPRIS
WEST
EDUCATIONAL SERVICES

4093 Specialty Place • Longmont, Colorado 80504 • (800) 547-6747
www.sopriswest.com

212STU/11-03/BAN/20M/121

TABLE OF CONTENTS
Developing Number Sense

UNIT 1	**Addition**	
	Working With Data	
Lesson 1	Place Value in Whole Numbers	
	Common Errors in Place Value	1
Lesson 2	Writing Numbers in Expanded Form	
	Reading Problems Carefully	3
Lesson 3	Basic and Extended Addition Facts	
	Problem Solving: What Is the Problem Asking For?	5
Lesson 4	Vertical and Horizontal Form	
	Problem Solving: Reading Carefully	7
Lesson 5	Writing Problems in Expanded Form	
	Problem Solving: More Practice Reading Carefully	9
Lesson 6	Expanded Addition	
	Using Graphs to Display Data	11
Lesson 7	Place Value Blocks	
	Reading Bar Graphs	13
Lesson 8	More Expanded Addition	
	Analyzing Data in Graphs	15
Lesson 9	Regrouping Using Expanded Addition	
	Problem Solving: Finding Important Information	17
Lesson 10	Comparing Methods of Regrouping	
	The Calculator: A Valuable Tool	19
Lesson 11	More Regrouping in Addition	
	Numbers on the Number Line	21
Lesson 12	Comparing Methods of Regrouping Again	
	Introducing the Blockheads	23
Lesson 13	Rounding Numbers: The Front-End Strategy	
	Using Rounded Numbers in Graphs	25
Lesson 14	Checking Your Work and Finding Errors	
	Horizontal Bar Graphs	29

Lesson 15	Review of Lessons 1 to 14	31
Lesson 16	Horizontal Expanded Addition *Reading Bar Graphs*	35
Lesson 17	More Horizontal Expanded Addition *Commuting in Addition*	37
Lesson 18	Horizontal Addition Using Place Value Blocks *Collecting Data and Constructing Graphs*	39
Lesson 19	Finishing Up Horizontal Expanded Addition *Pictographs*	41
Lesson 20	Unit 1 Review	43
UNIT 2	**Subtraction** ***Working With Data***	
Lesson 21	Fact Families in Subtraction *Fact-Tac-Toe*	47
Lesson 22	Expanded Subtraction *Reading Data From a Table*	49
Lesson 23	More Expanded Subtraction *Front-End Rounding in Subtraction*	51
Lesson 24	Regrouping With Expanded Subtraction *Number Lines With Different Scales*	53
Lesson 25	Comparing Two Methods of Regrouping in Subtraction *Quarter Rounding Strategy*	55
Lesson 26	More Practice Regrouping in Subtraction *Comparing Rounding Strategies*	59
Lesson 27	Comparing Regrouping Methods *More Practice Reading Data From Tables*	61
Lesson 28	Review of Lessons 21 to 27	63
Lesson 29	Identifying Common Subtraction Errors *The Blockheads Concert Tour: Florida*	65
Lesson 30	More on Common Errors *Double-Check the Bill*	67
Lesson 31	The Trouble With Zeros in Subtraction *Problem Solving: What's the Problem Asking For?*	69
Lesson 32	Why We Learn to Estimate *The Blockheads Concert Tour: On to Texas*	71
Lesson 33	Using Good Number Sense *The Bottom Line: Did We Make Any Money?*	73

Lesson 34 Writing About What You Have Learned
How Much Does a Movie Make? 75

Lesson 35 Unit 2 Review 77

UNIT 3 **Multiplication**
Measuring One-Dimensional Objects

Lesson 36 Basic and Extended Facts
Commuting in Multiplication 81

Lesson 37 Finding the 10 in Numbers
Estimating Distances 83

Lesson 38 Powers of 10
Making a Measuring Device 85

Lesson 39 Expanded Multiplication
Making a Ruler of 1s and 10s 87

Lesson 40 The Metric System
Measuring With a Metric Ruler 89

Lesson 41 Review of Lessons 36 to 40 93

Lesson 42 Comparing Methods of Regrouping
The World of Design 95

Lesson 43 Rounding Strategies in Multiplication
Making a Logo 97

Lesson 44 Why We Make Mistakes in Multiplication
Estimation Tasks 101

Lesson 45 Layout and Design
Designing a Web Page 105

Lesson 46 Estimation in Multiplication
Problem Solving: Finding Important Information 107

Lesson 47 Estimating and Using Calculators
Designing a Logo for the Prism Company 109

Lesson 48 Approximating Large Numbers
Scale Drawings 111

Lesson 49 Putting It All Together
How Many Times Bigger? 113

Lesson 50 Unit 3 Review 117

Contents v

UNIT 4   **Division**
Measuring Two-Dimensional Objects

Lesson 51   Multiplication and Division Fact Families
Measuring in Square Units   123

Lesson 52   Basic and Extended Division Facts
Comparing Sizes of Shapes   127

Lesson 53   Basic Division Facts on a Number Line
Profits From May Concerts   131

Lesson 54   Extended Division Facts on a Number Line
Same Shape, Same Size?   133

Lesson 55   Remainders
Estimating Square Units on a Map   137

Lesson 56   Using Calculators for Division
Rounding Strategies in Division   141

Lesson 57   "Near Fact" Division
Finding Area By Counting Squares   143

Lesson 58   Review of Lessons 51 to 57   147

Lesson 59   Division: The Traditional Algorithm
Careers in Architecture   151

Lesson 60   Comparing Division Algorithms
Estimating Areas in Floor Plans   155

Lesson 61   Looking at Bigger Division Problems
More Floor Plans   159

Lesson 62   Pulling Out the 10s
Problem Solving: The Arena Building   163

Lesson 63   Identifying Mistakes in Division on the Calculator
Problem Solving: Making Division Word Problems   165

Lesson 64   More Calculator Mistakes
Square Units and Triangular Units   167

Lesson 65   Unit 4 Review   169

UNIT 5   **Factors, Patterns, and Multiples**
Informal Geometry

Lesson 66   Arrays of Numbers 1 to 15
Review: Basic Operations and Bar Graphs   173

Lesson 67   Arrays of Numbers 16 to 25
Patterns of Numbers   175

Lesson 68	From Arrays to Factors *Estimating the Size of Irregularly Shaped Objects*	179
Lesson 69	Factor Rainbows *Areas of Rectangles and Squares*	183
Lesson 70	Writing Factor Lists *Reviewing Whole Number Operations*	187
Lesson 71	Composite and Prime Numbers *Perimeter*	189
Lesson 72	More on Primes *Numbers Between 1 to 100*	193
Lesson 73	Area *Different Shapes*	195
Lesson 74	Prime Factor Trees *Areas and Floor Plans*	197
Lesson 75	More Practice on Prime Factor Trees *Comparing Area and Perimeter*	201
Lesson 76	Area and Perimeter *Which Is Bigger?*	205
Lesson 77	Looking for Patterns in Area and Perimeter *Comparing Both Measures*	209
Lesson 78	Review of Lessons 66 to 77	213
Lesson 79	Dividing Rules *Measuring Area*	217
Lesson 80	More Practice With Dividing Rules *Geometry*	221
Lesson 81	Finding Prime Factors for Large Numbers *More Geometry*	225
Lesson 82	Finding Common Factors *More Patterns of Numbers*	229
Lesson 83	More Practice Finding Common Factors *Geometry and Art*	231
Lesson 84	Greatest Common Factor (GCF) *Looking at Patterns in Art*	233
Lesson 85	More Practice With GCF *Congruent Shapes*	235
Lesson 86	Making Patterns *Review of Prime Factor Trees*	237

Lesson 87	Challenge Lesson: Strategies for Finding GCF *Using Prime Factor Trees*	241
Lesson 88	Review of Lessons 79 to 87	243
Lesson 89	Patterns of Numbers: Evens and Odds *Expanding and Contracting Objects*	249
Lesson 90	Patterns of Numbers *Square Numbers*	253
Lesson 91	Patterns of Numbers: Square Numbers and Odd Numbers *Expanding Triangles*	257
Lesson 92	Patterns of Numbers: Triangular Numbers *Problem Solving: Finding Patterns*	261
Lesson 93	Patterns of Numbers: When Numbers Repeat *Scaling: Expanding and Contracting*	263
Lesson 94	Common Multiples *Changing Objects*	267
Lesson 95	Finding the Least Common Multiple (LCM) *Symmetry*	271
Lesson 96	More Practice Finding LCM *Wheels Inside a Cuckoo Clock*	275
Lesson 97	Mobiles—Moving Art *Practice With LCM*	279
Lesson 98	Adding a Third Wheel to the Cuckoo Clock *Rotational Symmetry*	283
Lesson 99	Number Sense *Summing Up*	287
Lesson 100	Unit 5 Review	291

PLACE VALUE IN WHOLE NUMBERS
Common Errors in Place Value

Name _____

PART A
REVIEW OF PLACE VALUE

Activity 1: Rewrite the following numbers by placing the digits in the appropriate places in the place value charts below. Use the example as a guide.

587,658   EXAMPLE

100 Millions	10 Millions	Millions	100 Thousands	10 Thousands	Thousands	100s	10s	1s
			5	8	7	6	5	8

1. 6,421,359

100 Millions	10 Millions	Millions	100 Thousands	10 Thousands	Thousands	100s	10s	1s

2. 500,407

100 Millions	10 Millions	Millions	100 Thousands	10 Thousands	Thousands	100s	10s	1s

3. 21,058

100 Millions	10 Millions	Millions	100 Thousands	10 Thousands	Thousands	100s	10s	1s

Student Workbook • Number Sense   Unit 1 Addition   1

4. 105,009,670

100 Millions	10 Millions	Millions	100 Thousands	10 Thousands	Thousands	100s	10s	1s

Activity 2: Write the value of the bold, underlined digit.

EXAMPLE   2**7**4,145 _____ 70,000 _____

1. 348,**5**97 _____

2. **1**23,405 _____

3. 50**8**,112,485 _____

4. 3**0**0,001 _____

5. 5,4**7**3 _____

Activity 3: The following numbers are written using English words. Tell how many digits each number will have and rewrite the number using digits.

The Number Written in Words	How Many Digits?	Write the Number
Five hundred thousand seven hundred		
Four thousand sixty-five		
Eight hundred one		
Nine million forty thousand seventy		
Fifty-four		

WRITING NUMBERS IN EXPANDED FORM
Reading Problems Carefully

Lesson 2

Name _____

PART A
WARM-UP ACTIVITY: REVIEW OF PLACE VALUE

Activity: Answer the following questions about place value for the number 896,403,157.

1. What is the value of the digit "6"? _____
2. What digit is in the thousands place? _____
3. What is the value of the digit "5"? _____
4. What digit is in the 10 millions place? _____
5. What is the value of the digit "7"? _____
6. What place is the zero in? _____

PART B
EXPANDED FORM OF NUMBERS

Activity 1: Practice changing numbers to *expanded form*.

1. 543 → _____
2. 605 → _____
3. 890 → _____
4. 1,005 → _____
5. 2,012 → _____
6. 5,396 → _____

Activity 2: Practice changing numbers from *expanded form* to the regular way of writing the numbers.

1. 900 + 70 + 2 → _____
2. 700 + 20 + 0 → _____
3. 400 + 0 + 1 → _____
4. 5,000 + 0 + 70 + 2 → _____
5. 7,000 + 300 + 0 + 1 → _____
6. 8,000 + 10 + 3 → _____

Student Workbook • *Number Sense*   Unit 1 Addition

PART C
WORD PROBLEMS

Practice *reading carefully* and identifying the *important information* in the following word problems. Do not solve.

Problem 1: There were 5,200 hot dogs sold at the ballpark last night. The attendance was 14,507. The hot dogs cost $3.75 each. The night before last, there were 4,100 hot dogs sold. The owner of the ballpark wanted to know how many hot dogs were sold at these last 2 games. The attendance that night was 17,129.

What's the problem asking for? _____

_____

Underline the important information.

Problem 2: Sean met 5 friends at the arcade. He had $20 to spend. He spent $3 on air hockey. His friends each played 6 games. Sean spent $10 on video games. Then he spent $2 for a drink and a snack. He earned 175 tickets from the video games. His friend Kevin earned 200 tickets. Sean's mother asked him how much he spent at the arcade. What was his answer?

What's the problem asking for? _____

_____

Underline the important information.

BASIC AND EXTENDED ADDITION FACTS

Problem Solving:
What Is the Problem Asking For?

Name _____

Lesson 3

PART A
WARM-UP ACTIVITY: SOLVING BASIC FACTS

Activity: Solve the following basic facts.

1. 4 + 9 = ____
2. 7 + 8 = ____
3. 9 + 4 = ____
4. 8 + 7 = ____

5. 6 + 5 = ____
6. 7 + 9 = ____
7. 5 + 6 = ____
8. 9 + 7 = ____

9. 6 + 8 = ____
10. 8 + 9 = ____
11. 8 + 6 = ____
12. 9 + 8 = ____

PART B
SOLVING BASIC AND EXTENDED FACTS

Activity 1: Solve the following basic and extended facts.

1. 4 + 9 = ____   40 + 90 = ____   400 + 900 = ____
2. 7 + 8 = ____   70 + 80 = ____   700 + 800 = ____
3. 6 + 5 = ____   60 + 50 = ____   600 + 500 = ____
4. 7 + 9 = ____   70 + 90 = ____   700 + 900 = ____
5. 6 + 8 = ____   60 + 80 = ____   600 + 800 = ____
6. 8 + 9 = ____   80 + 90 = ____   800 + 900 = ____

Activity 2: Write the extended fact equation for the following word problem and then state the basic fact.

Problem: The snack bar at Marshall Field made $80 from the sale of hot dogs and $50 from the sale of popcorn at the game last Saturday. What was the total amount sold for the 2 items?

What is the extended fact? _____

What is the basic fact? _____

Student Workbook • *Number Sense*   Unit 1 Addition   5

PART C
WORD PROBLEMS

Activity: Practice *reading carefully* and identifying *important information* in the following word problems. Do not solve.

Problem 1: Nick has 25 CDs. They each cost about $15. His CD player cost $60. He likes only one song on each CD. He could buy CDs of the singles of the songs he likes for about $5 each. How much money would he have saved had he bought 25 CDs of singles instead?

What's the problem asking for? _____

_____

Underline the important information.

Problem 2: Jared likes to play arcade games at the mall. He collects the tickets that he earns from playing the games and saves them. He currently has 2,500 tickets. If he earns 500 tickets each time he visits the arcade, he wants to know how long it will take him to earn the big prize. Jared is saving for one of the big prizes that require 7,500 tickets. He spends about $10 each time he goes to the arcade.

What's the problem asking for? _____

_____

Underline the important information.

VERTICAL AND HORIZONTAL FORM
Problem Solving: Reading Carefully

Lesson 4

Name _____

PART A
WARM-UP ACTIVITY: BASIC AND EXTENDED FACTS

Activity: Solve the following basic and extended facts.

1. 9 + 8 = ____   90 + 80 = ____   900 + 800 = ____
2. 6 + 7 = ____   60 + 70 = ____   600 + 700 = ____
3. 9 + 4 = ____   90 + 40 = ____   900 + 400 = ____
4. 5 + 8 = ____   50 + 80 = ____   500 + 800 = ____

PART B
VERTICAL AND HORIZONTAL FORM

Activity 1: Rewrite the following equations in vertical form. Do not solve.

1. 37 + 42 → + _____

2. 29 + 95 → + _____

3. 58 + 27 → + _____

4. 107 + 39 → + _____

Activity 2: Rewrite the following equations in horizontal form. Do not solve.

1. 68
   + 42   → _____ + _____

2. 37
   + 85   → _____ + _____

3. 59
   + 11   → _____ + _____

4. 120
   + 307   → _____ + _____

Student Workbook • *Number Sense* Unit 1 Addition

PART C
WORD PROBLEMS

Read the problem and answer the questions below.

1. The Blue Jays hit 24 home runs in June and 28 home runs in July. In August, the team's pitcher, Emilio Sanchez, had 2 games where he struck out 15 batters. In August, the Blue Jays hit 21 home runs. The coach wants to know how many home runs the Blue Jays hit in June and July. The team is getting ready for 4 straight games at home.

   What's the problem asking for? _____

   _____

   Underline the important information.

2. It takes 5 hours to fly from San Francisco to New York. It takes another 7 hours to fly from New York to Rome, Italy. Many people travel this route. They usually stay in New York at a hotel by the airport and go to Rome the next day. Andrea wants to travel from San Francisco to New York to Rome. She wants to know how long the plane flights are, all together, but she does not want to spend the night in New York. She will be on a business trip, and she won't have a lot of extra time.

   What's the problem asking for? _____

   _____

   Underline the important information.

WRITING PROBLEMS IN EXPANDED FORM

Problem Solving: More Practice Reading Carefully

Lesson 5

Name _____

PART A
WARM-UP ACTIVITY: EXPANDED FORM

Activity: Write the following problems in expanded form. Do not solve.

1. 407 → _____
2. 312 → _____
3. 650 → _____
4. 1,309 → _____
5. 2,002 → _____
6. 9,096 → _____

PART B
NUMBERS IN EXPANDED AND REGULAR FORM

Activity 1: Write the following problems in expanded form. Do not solve.

EXAMPLE

$$\begin{array}{r} 24 \\ + 83 \end{array} \rightarrow \begin{array}{r|r} 20 & 4 \\ + 80 & 3 \end{array}$$

1. $\begin{array}{r} 37 \\ + 42 \end{array}$ →

2. $\begin{array}{r} 29 \\ + 95 \end{array}$ →

3. $\begin{array}{r} 58 \\ + 27 \end{array}$ →

Student Workbook • Number Sense   Unit 1 Addition

Activity 2: Write the following problems the regular way. Do not solve.

EXAMPLE

```
  30 | 7          37
+ 50 | 4    →   + 54
```

1.
```
  60 | 8
+ 40 | 2    →   + _____
```

2.
```
  30 | 7
+ 80 | 5    →   + _____
```

3.
```
  50 | 9
+ 10 | 1    →   + _____
```

4.
```
  100 | 60 | 5
+     | 80 | 8   →   + _____
```

5.
```
  300 | 0 | 7
+ 500 | 0 | 0   →   + _____
```

PART C
WORD PROBLEMS

For each word problem, do the following:

- Decide what the problem is asking for.
- Write a sentence that tells what the problem is asking for.
- Underline the important information.

1. Leroy and Jamaal went to the movies on Saturday afternoon. Each ticket cost $5.50, and they bought snacks. Sarah and Julia were at the movies that afternoon, too. Leroy bought popcorn and a drink for $5 and Jamaal just bought a drink for $3. When Leroy got home, his mom wanted to know how much he and Jamaal spent at the movies. Leroy's mom had just gotten home from work.

2. Climbing a tall mountain can take several days and a lot of planning. A group of climbers from Kenya wanted to climb one of the tallest mountains in the world. They hiked up the mountain 2 days before they reached the first base camp. It took them another 2 days to reach the second base camp. It was very cold that morning. Many of the climbers took out jackets they didn't think they would need. Finally, they hiked that day to the top of the mountain. How many days did it take them to hike to the top of the mountain?

EXPANDED ADDITION
Using Graphs to Display Data

Name _____

PART A
WARM-UP ACTIVITY: BASIC AND EXTENDED FACTS

Activity: Solve the following basic and extended facts.

1. 90 + 40 = ____
2. 7 + 8 = ____
3. 900 + 400 = ____
4. 9 + 4 = ____

5. 6 + 5 = ____
6. 70 + 80 = ____
7. 700 + 800 = ____
8. 800 + 900 = ____

9. 60 + 50 = ____
10. 8 + 9 = ____
11. 80 + 90 = ____
12. 600 + 500 = ____

PART B
EXPANDED ADDITION

Activity: Rewrite the following equations in expanded form. Then write the answer to the right.

1. 37 →
   + 42

   + _____ | _____   Answer ____

2. 61 →
   + 25

   + _____ | _____   Answer ____

3. 74 →
   + 13

   + _____ | _____   Answer ____

Student Workbook • *Number Sense*   Unit 1 Addition   **11**

PLACE VALUE BLOCKS
Reading Bar Graphs

Name _____

Lesson 7

PART A
WARM-UP ACTIVITY: BASIC AND EXTENDED FACTS

Activity: Solve the following basic and extended facts.

1. 4 + 9 = ____
2. 9 + 4 = ____
3. 7 + 8 = ____
4. 8 + 7 = ____

5. 60 + 50 = ____
6. 50 + 60 = ____
7. 7 + 9 = ____
8. 9 + 7 = ____

9. 6 + 8 = ____
10. 8 + 6 = ____
11. 80 + 90 = ____
12. 90 + 80 = ____

PART B
EXPANDED ADDITION

Activity 1: Rewrite the following equations in expanded form and solve. Combine the answer and write it on the line to the right.

1. 337 →
   + 542

   +

   Answer ____

2. 461 →
   + 225

   +

   Answer ____

3. 780 →
   + 109

   +

   Answer ____

Student Workbook • *Number Sense*   Unit 1 Addition   **13**

Activity 2: In these problems, numbers are written using the following notation, which represents place value blocks. Rewrite the numbers using digits.

☐ = Hundreds   ▯ = Tens   • = Ones

EXAMPLE

The number is 136.

1.

The number is ____

2.

The number is ____

3.

The number is ____

14   Unit 1 Addition • Lesson 7   Student Workbook • *Number Sense*

MORE EXPANDED ADDITION
Analyzing Data in Graphs

Lesson 8

Name _____

PART A
WARM-UP ACTIVITY: EXTENDED FACTS

Activity: Solve the following extended facts.

1. 500 + 400 = ____
2. 40 + 50 = ____
3. 300 + 700 = ____
4. 90 + 80 = ____

5. 80 + 30 = ____
6. 300 + 800 = ____
7. 60 + 20 = ____
8. 200 + 600 = ____

9. 400 + 500 = ____
10. 50 + 40 = ____
11. 900 + 800 = ____
12. 80 + 90 = ____

PART B
EXPANDED ADDITION

Activity 1: Rewrite the following equations in expanded form and solve. Combine the answer and write it on the line to the right.

1.   312 →   Answer ____
   + 457

2.   207 →   Answer ____
   + 691

3.   566 →   Answer ____
   + 303

Student Workbook • Number Sense   Unit 1 Addition   15

Activity 2: Rewrite the following equations using place value block notation. Do not solve the problems, just write the problems using block notation. Remember, that is:

☐ = Hundreds   ▯ = Tens   • = Ones

1. 748
   + 101   Rewrite the equation here:

2. 612
   + 276   Rewrite the equation here:

3. 503
   + 416   Rewrite the equation here:

Activity 3: The following equations are written using place value block notation. Rewrite using numbers and then solve.

1. ☐☐☐☐☐☐ | ▯▯▯▯ | •• •••
   +   ☐☐☐

2. ☐☐☐ | | •••
   +   ☐☐ | ▯ | ••

3. ☐☐☐ | ▯▯ | •• •••• •
   + ☐☐☐☐ | ▯ | ••

16   Unit 1 Addition • Lesson 8   Student Workbook • *Number Sense*

REGROUPING USING EXPANDED ADDITION
Problem Solving: Finding Important Information

LESSON 9

Name _____

PART A
WARM-UP ACTIVITY: BASIC AND EXTENDED FACTS

Activity: Solve the following basic and extended facts.

1. 9 + ____ = 17   90 + ____ = 170   900 + ____ = 1,700
2. 8 + ____ = 17   80 + ____ = 170   800 + ____ = 1,700
3. ____ + 4 = 13   ____ + 40 = 130   ____ + 400 = 1,300
4. ____ + 9 = 13   ____ + 90 = 130   ____ + 900 = 1,300

PART B
EXPANDED ADDITION WITH REGROUPING

Activity: Rewrite the following problems in expanded form. Solve.

1. 57 →
   + 28

   +

   Answer ____

2. 45 →
   + 16

   +

   Answer ____

Student Workbook • *Number Sense*   Unit 1 Addition   17

PART C
PROBLEM SOLVING

For each word problem, do the following:

- Decide what the problem is asking for.
- Write a sentence that tells what the problem is asking for.
- Underline the important information.

1. This list shows the top 5 movies and how much they made over 1 weekend.

Movies	Weekend Sales
Harry Potter	$82 million
Monsters, Inc.	$32 million
Spy Game	$30 million
Black Knight	$15 million
Shallow Hal	$12 million

One movie company was interested in how much more money the top movie made than the number 2 movie. This company was thinking about making a movie similar to *Harry Potter*. The sales amounts listed in the chart are for just 1 weekend. The sales amounts do not include the total money the movies have made since they were released.

2. The movie that was tenth that weekend was *Life as a House*. It made $3 million dollars over the weekend. The movie had been out for 5 weeks, and before that weekend it had made a total of $9 million. The producer wanted to know how much money the movie had made since it first came out, including the $3 million weekend. She was planning to use this number to figure out when the movie would make a profit.

COMPARING METHODS OF REGROUPING
The Calculator: A Valuable Tool

LESSON 10

Name _____

PART A
WARM-UP ACTIVITY: BASIC AND EXTENDED FACTS

Activity: Solve the following basic and extended facts.

1. 2 + 9 = ____   90 + 20 = ____   200 + 900 = ____
2. 3 + 7 = ____   70 + 30 = ____   300 + 700 = ____
3. ____ + 8 = 13   ____ + 80 = 130   ____ + 800 = 1,300
4. ____ + 5 = 13   ____ + 50 = 130   ____ + 500 = 1,300

PART B
EXPANDED ADDITION AND TRADITIONAL ADDITION

Activity 1: Rewrite the following problems in expanded form and then solve.

1. 47 →
   + 35

   Answer ____

2. 29 →
   + 18

   Answer ____

Activity 2: Now solve the two problems from Activity 1 using the traditional addition algorithm. Check your answers with your calculator.

$$\begin{array}{r} 47 \\ + 35 \\ \hline \end{array} \qquad \begin{array}{r} 29 \\ + 18 \\ \hline \end{array}$$

Student Workbook • Number Sense   Unit 1 Addition   19

Activity 3: Compare the traditional method with the expanded method for addition. Write what's the same and what's different between the two methods for addition.

_____

_____

_____

_____

_____

_____

_____

_____

_____

_____

_____

_____

_____

_____

_____

_____

PART C
USE YOUR CALCULATOR

1. 5,976
   + 398

2. 62,495
   + 28,786

3. 1,323,597
   + 565,318

MORE REGROUPING IN ADDITION
Numbers on the Number Line

LESSON 11

Name _____

PART A
WARM-UP ACTIVITY: BASIC FACTS

Activity: Solve the following basic facts.

1. 6 + ____ = 12
2. ____ + 6 = 12
3. 4 + 8 = ____
4. ____ = 8 + 4

5. ____ + 6 = 13
6. 13 = ____ + 6
7. 12 = ____ + 8
8. 4 + ____ = 12

9. 17 = 8 + ____
10. 9 + ____ = 17
11. 9 + 9 = ____
12. ____ = 9 + 9

PART B
EXPANDED ADDITION AND USING A CALCULATOR

Activity 1: Solve the following equations first using expanded form and then using the traditional method of addition.

Expanded Form

1. 39 →   +_____   Answer ____
   + 48

2. 24 →   +_____   Answer ____
   + 58

3. 76 →   +_____   Answer ____
   + 17

Student Workbook • *Number Sense*   Unit 1 Addition   21

Traditional Method

```
  39          24          76
+ 48        + 58        + 17
```

Activity 2: The following problems show why we need calculators. Find the answers to these problems using a calculator.

```
  554       2,987       13,784
+  87     + 6,983     +  7,228
```

PART C
MAKING WORD PROBLEMS

Activity: We've been trying to get you to look closely at word problems. Here's something new. The paragraph below *does not* ask you to do anything. Instead, *you* are to turn it into a word problem. Read the paragraph and do the following:

- Below the paragraph, write a question so that this becomes a word problem.
- Underline the information that is important to answering the question.
- Write the answer to the question.

Radio stations make a lot of money from companies buying advertising time on the radio. Some radio stations make millions of dollars each year on advertising. KWWQ in Sacramento made $2 million on advertising in 1 year. KWWQ plays a lot of hard rock. KRQZ is in the same city. It plays classical music. It made $650,000 on advertising during the same year. Finally, KRRA plays country western music in the Sacramento area. It made $4 million on advertising.

Write a question that makes this paragraph a word problem and then write an answer to the question.

Question _____

Answer _____

LESSON 12

COMPARING METHODS OF REGROUPING AGAIN
Introducing the Blockheads

Name _____

PART A
WARM-UP ACTIVITY: BASIC AND EXTENDED FACTS

Activity: Solve the following basic and extended facts.

1. 4 + 9 = ____
2. ____ = 40 + 90
3. 900 + 400 = ____
4. 6 + ____ = 13

5. 130 = ____ + 60
6. ____ + 600 = 1,300
7. ____ + 8 = 12
8. 120 = ____ + 80

9. 1,200 = 800 + ____
10. 9 + ____ = 14
11. 90 + ____ = 140
12. 1,400 = ____ + 900

PART B
EXPANDED ADDITION

Activity: Solve the following equations first using expanded form and then using the traditional method of addition.

Expanded Form

1. 239 →   Answer ____
   + 25

2. 576 →   Answer ____
   + 47

3. 179 →   Answer ____
   + 724

Traditional Method

   239   576   179
  + 25   + 47   + 724

Student Workbook • Number Sense   Unit 1 Addition   23

PART C
READING GRAPHS

Activity: Look at the graph. Then do the following for the two problems that follow:

- Finish the problem by writing a question.
- Answer the question.

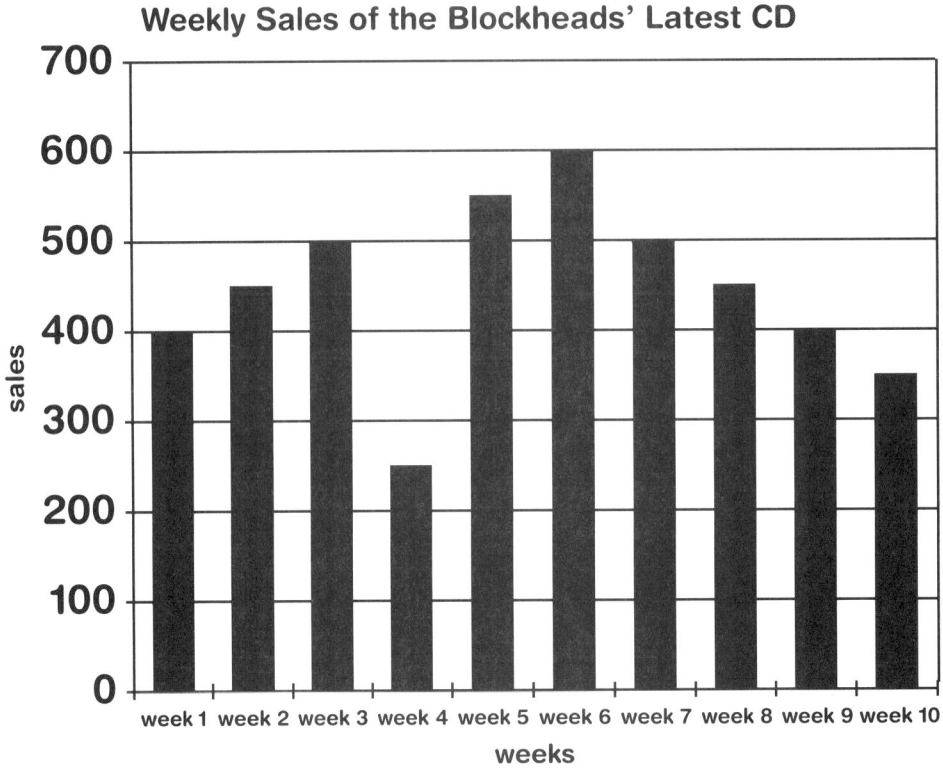

1. Cyndee has been looking over the graph of weekly sales of the Blockheads' latest CD. Cyndee notices that the top sales were in Week 6. She also notices that the lowest sales were in Week 4.

    Question _____

    Answer _____

2. Cyndee notices that there is a trend in sales over the last 5 weeks. For every week, sales are less than for the week before. The company that sells the CDs has a rule that it stops advertising when the sales get below $200 a week.

    Question _____

    Answer _____

ROUNDING NUMBERS: THE FRONT-END STRATEGY
Using Rounded Numbers in Graphs

LESSON 13

Name _____

PART A
WARM-UP ACTIVITY: EXTENDED FACTS

Activity: Solve the following extended facts.

1. 700 + 900 = ____
2. 70 + 50 = ____
3. 50 + ____ = 130
4. 1,300 = 500 + ____
5. 600 + ____ = 1300
6. 130 = 60 + ____
7. ____ + 600 = 1,300
8. 900 + 700 = ____
9. ____ + 500 = 1,300
10. 80 + 40 = ____
11. 400 + 800 = ____
12. ____ = 40 + 80

PART B
FRONT-END ROUNDING

Activity: Practice rounding the following numbers using the front-end strategy, then solve the extended fact. Find the exact answer on your calculator and compare.

1. 49 + 82 = ____   Round 49 to ____   Round 82 to ____

   The new equation is the extended fact ____ + ____ = ____ .

   Calculator answer: 49 + 82 = ____

2. 92 + 69 = ____   Round 92 to ____   Round 69 to ____

   The new equation is the extended fact ____ + ____ = ____ .

   Calculator answer: 92 + 69 = ____

3. 697 + 921 = ____   Round 697 to ____   Round 921 to ____

   The new equation is the extended fact ____ + ____ = ____ .

   Calculator answer: 697 + 921 = ____

4. 401 + 837 = ____   Round 401 to ____   Round 837 to ____

   The new equation is the extended fact ____ + ____ = ____ .

   Calculator answer: 401 + 837 = ____

PART C
EXPANDED ADDITION

Activity: Solve the following equations first using expanded form and then using the traditional method of addition. Check your answers with your calculator.

Expanded Form

1. 176 →
    + 27

    + | | Answer ____

2. 578 →
    + 13

    + | | Answer ____

3. 149 →
    + 725

    + | | Answer ____

4. 208 →
    + 394

    + | | Answer ____

Traditional Method

| 176 | 578 | 149 | 208 |
| + 27 | + 13 | + 725 | + 394 |

26 Unit 1 Addition • Lesson 13   Student Workbook • *Number Sense*

PART D
PROBLEM SOLVING WITH ROUNDING

Activity: Solve the following word problem by rounding the numbers using the front-end strategy and solving the extended fact equation. Then use your calculator to find the exact answer and compare.

Problem 1: There are 577 students in the 10th grade at Central High, and there are 621 students in the 11th grade. How many students are in both grades combined?

Extended Fact Equation

(rounding): ____ + ____ = ____

Actual Equation

(exact numbers): ____ + ____ = ____

Comparison—What is the difference between the extended fact answer and the exact answer?

Problem 2: Central High is trying to figure out how many people buy lunch during 2nd and 3rd periods. Yesterday, 142 students bought lunch during 2nd period, and 135 students bought lunch during 3rd period. How many students bought lunch for 2nd and 3rd period combined?

Extended Fact Equation

(rounding): ____ + ____ = ____

Actual Equation

(exact numbers): ____ + ____ = ____

Comparison—What is the difference between the extended fact answer and the exact answer?

CHECKING YOUR WORK AND FINDING ERRORS
Horizontal Bar Graphs

Lesson 14

Name _____

PART A
WARM-UP ACTIVITY: ROUNDING TO EXTENDED FACTS

Activity: Round the following numbers and solve the extended fact.

1. 27 + 32 ⟶ Rounds to the extended fact ____ + ____ = ____
2. 46 + 11 ⟶ Rounds to the extended fact ____ + ____ = ____
3. 401 + 301 ⟶ Rounds to the extended fact ____ + ____ = ____
4. 599 + 839 ⟶ Rounds to the extended fact ____ + ____ = ____

PART B
FINDING THE ERRORS

Activity: Find which of the following problems have been answered correctly and which have been answered incorrectly. Use expanded form to help you fix the error and find the correct answer.

```
  ¹                ¹               ¹
  29               67              18
+ 87             + 93            + 72
─────           ─────           ─────
 106             160              90
```

Work Area:

Student Workbook • Number Sense — Unit 1 Addition

REVIEW OF LESSONS 1 TO 14

Lesson 15

Name _____

PART A
WARM-UP ACTIVITY: BASIC AND EXTENDED FACTS

Activity: Solve the following basic and extended facts.

1. 9 + ____ = 15
2. ____ + 9 = 15
3. 150 = ____ + 90
4. 1,500 = 900 + ____

5. 8 + ____ = 13
6. 13 = 8 + ____
7. ____ + 800 = 1,300
8. 80 + ____ = 130

9. ____ + 4 = 13
10. 13 = 4 + ____
11. 400 + ____ = 1,300
12. 40 + ____ = 130

PART B
EXPANDED ADDITION

Activity: Solve the following equations first using expanded form and then using the traditional method of addition. Check your answers with your calculator.

Expanded Form

1. 39
   + 24   →   + _____|_____   Answer ____

2. 179
   + 56   →   + _____|_____|_____   Answer ____

3. 895
   + 438   →   + _____|_____|_____   Answer ____

Student Workbook • *Number Sense* — Unit 1 Addition

Traditional Method

```
   39          179          895
 + 24        +  56        + 438
```

PART C
FRONT-END ROUNDING

Activity 1: Practice rounding the following numbers using the front-end strategy, then solve the extended fact. Find the exact answer on your calculator and compare.

1. 79 + 92 = ____

   Round 79 to ____

   Round 92 to ____

   The new equation is the extended fact ____ + ____ = ____

   Calculator answer: ____

2. 642 + 721 = ____

   Round 642 to ____

   Round 721 to ____

   The new equation is the extended fact ____ + ____ = ____

   Calculator answer: ____

Activity 2: Solve the following word problems by rounding the numbers using the front-end strategy and solving the extended fact equation. Then use your calculator to find the exact answer and compare.

> **Problem 1:** ABC Records sold 498 CDs in its first month of business and 631 CDs in its second month of business. How many CDs did the company sell in the 2 months?
>
> Extended Fact (rounding) Equation ____ + ____ = ____
>
> Actual Equation ____ + ____ = ____
>
> What is the difference between the estimate and the exact answer?

Problem 2: The record for the most number of baseball games won in a regular season was set in 1906. The Seattle Mariners broke that record 95 years later. In what year did the Mariner's break the record?

Extended Fact Equation ____ + ____ = ____

Actual Equation ____ + ____ = ____

What is the difference between the estimate and the exact answer?

PART D
FINDING ERRORS

Activity: Determine which of the following problems have been answered correctly and which have been answered incorrectly. Use expanded form to help you find and fix the errors. Check with your calculator.

$$\begin{array}{r} \overset{1}{3}9 \\ +\ 88 \\ \hline 117 \end{array} \qquad \begin{array}{r} \overset{1}{1}18 \\ +234 \\ \hline 452 \end{array} \qquad \begin{array}{r} \overset{11}{2}98 \\ +\ 372 \\ \hline 670 \end{array}$$

Work Area:

HORIZONTAL EXPANDED ADDITION
Reading Bar Graphs

LESSON 16

Name _____

PART A
WARM-UP ACTIVITY: EXTENDED FACTS

Activity: Solve these extended facts.

1. 60 + ____ = 110
2. ____ + 60 = 110
3. 40 + 90 = ____
4. ____ = 90 + 40

5. ____ + 50 = 120
6. 120 = ____ + 50
7. 130 = ____ + 80
8. 80 + ____ = 130

9. 170 = 90 + ____
10. 90 + ____ = 170
11. 90 + 90 = ____
12. ____ = 90 + 90

PART B
HORIZONTAL ADDITION WITH COMMUTING

Activity 1: Practice using horizontal and expanded form to solve two-digit x two-digit addition problems.

Step 1: Rewrite in expanded form.

37 + 42 → ____ + ____ + ____ + ____

Step 2: Rearrange the place values.

37 + 42 → ____ + ____ + ____ + ____
                   Tens   Ones

Step 3: Combine and solve.

37 + 42 → ____ + ____
             Tens   Ones

Answer ____

Activity 2: Work these problems by commuting.

55 + 22 = _____
34 + 44 = _____

63 + 25 = _____
31 + 38 = _____

Student Workbook • *Number Sense*   Unit 1 Addition   35

MORE HORIZONTAL EXPANDED ADDITION
Commuting in Addition

LESSON 17

Name _____

PART A
WARM-UP ACTIVITY: BASIC AND EXTENDED FACTS

Activity 1: Solve these basic and extended facts.

1. 40 + 90 = ____
2. 800 + 400 = ____
3. 7 + 7 = ____
4. 6 + 7 = ____

5. 9 + 5 = ____
6. 5 + 7 = ____
7. 70 + 60 = ____
8. 90 + 40 = ____

9. 4 + 8 = ____
10. 800 + 800 = ____
11. 500 + 900 = ____
12. 7 + 5 = ____

Activity 2: Be careful solving these problems. They're a mix of facts and extended addition.

1. 40 + 9 = ____
2. 5 + 4 = ____
3. 60 + 7 = ____
4. 90 + 70 = ____

5. 8 + 5 = ____
6. 50 + 7 = ____
7. 70 + 60 = ____
8. 9 + 40 = ____

9. 4 + 80 = ____
10. 7 + 80 = ____
11. 50 + 60 = ____
12. 70 + 8 = ____

PART B
HORIZONTAL ADDITION WITH COMMUTING

Activity 1: Rewrite the following equations in horizontal and expanded form and rearrange the places—tens with the tens, ones with the ones, etc. Then combine and solve.

1. 37 + 42 → ____ + ____ + ____ + ____

   ____ + ____   ____ + ____
   Tens   Ones

   ____   +   ____
   Tens   Ones

   Answer ____

Student Workbook • *Number Sense* — Unit 1 Addition — 37

2. 61 + 25 → ___ + ___   +   ___ + ___

   ___ + ___   +   ___ + ___
   Tens   *Ones*

   ___   +   ___
   Tens   *Ones*

   Answer ___

Activity 2: Use place value block notation to represent the following equations (using horizontal and expanded form) and then solve.

▯ = Tens   • = Ones

1. 12 + 11 =

2. 52 + 31 =

3. 43 + 21 =

4. 62 + 33 =

HORIZONTAL ADDITION USING PLACE VALUE BLOCKS
Collecting Data and Constructing Graphs

LESSON 18

Name _____

PART A
WARM-UP ACTIVITY: BASIC AND EXTENDED FACTS

Activity: Solve the following extended facts. Notice the turn-around facts.

1. 500 + 900 = ____
2. 90 + 50 = ____
3. 300 + 800 = ____
4. 80 + 30 = ____
5. 90 + 30 = ____
6. 300 + 900 = ____
7. 60 + 70 = ____
8. 700 + 600 = ____
9. 400 + 800 = ____
10. 80 + 40 = ____
11. 900 + 800 = ____
12. 80 + 90 = ____

PART B
HORIZONTAL ADDITION WITH COMMUTING

Activity 1: Rewrite the following equations in horizontal and expanded form and rearrange the places—tens with the tens, ones with the ones, etc. Then combine and solve.

1. 437 + 42 → ____ + ____ + ____ + ____ + ____

   ____ + ____ + ____ + ____ + ____
   Hundreds   Tens   Ones

   ____ + ____ + ____
   Hundreds Tens   Ones

   Answer ____

2. 123 + 410 → ____ + ____ + ____ + ____ + ____ + ____

   ____ + ____ + ____ + ____ + ____ + ____
   Hundreds   Tens   Ones

   ____ + ____ + ____
   Hundreds Tens   Ones

   Answer ____

Student Workbook • *Number Sense*   Unit 1 Addition   39

Activity 2: Use place value block notation to represent the following equations (using horizontal and expanded form) and then solve.

☐ = Hundreds   ▯ = Tens   • = Ones

1. 748 + 31 =

2. 612 + 206 =

3. 530 + 416 =

4. 300 + 291 =

FINISHING UP HORIZONTAL EXPANDED ADDITION
Pictographs

LESSON 19

Name _____

PART A
WARM-UP ACTIVITY: BASIC AND EXTENDED FACTS

Activity: Solve the following basic and extended facts.

1. 9 + ____ = 11   90 + ____ = 110   900 + ____ = 1,100

2. 8 + ____ = 14   80 + ____ = 140   800 + ____ = 1,400

3. ____ + 7 = 15   ____ + 70 = 150   ____ + 700 = 1,500

4. 5 + 6 = ____   50 + 60 = ____   600 + 500 = ____

PART B
HORIZONTAL ADDITION WITH COMMUTING

Activity: Rewrite the following problems in horizontal and expanded form, rearrange, combine, and then solve.

1. 57 + 21 → ____ + ____   +   ____ + ____

    ____ + ____   +   ____ + ____

    ____   +   ____

    Answer ____

2. 32 + 43 → ____ + ____   +   ____ + ____

    ____ + ____   +   ____ + ____

    ____   +   ____

    Answer ____

Student Workbook • *Number Sense* — Unit 1 Addition — 41

3. 123 + 45 ⟶ ___ + ___ + ___   +   ___ + ___

___ + ___ + ___ + ___ + ___

___ + ___ + ___

Answer ___

4. 234 + 51 ⟶ ___ + ___ + ___   +   ___ + ___

___ + ___ + ___ + ___ + ___

___ + ___ + ___

Answer ___

5. 116 + 321 ⟶ ___ + ___ + ___   +   ___ + ___ + ___

___ + ___ + ___ + ___ + ___ + ___

___ + ___ + ___

Answer ___

6. 207 + 442 ⟶ ___ + ___ + ___   +   ___ + ___ + ___

___ + ___ + ___ + ___ + ___ + ___

___ + ___ + ___

Answer ___

UNIT 1 REVIEW

LESSON 20

Name _____

PART A
WARM-UP ACTIVITY: BASIC AND EXTENDED FACTS

Activity: Solve the following basic and extended facts.

1. 9 + ____ = 15   90 + ____ = 150   900 + ____ = 1,500
2. 8 + ____ = 16   80 + ____ = 160   800 + ____ = 1,600
3. ____ + 7 = 12   ____ + 70 = 120   ____ + 700 = 1,200
4. 7 + 6 = ____   70 + 60 = ____   600 + 700 = ____

PART B
PLACE VALUE

Activity: Write the value of the bold, underlined digit in each of the following numbers.

1. 37,**5**10 _____
2. 4,0**1**7 _____
3. **6**,500 _____
4. 22**3** _____

PART C
EXPANDED FORM

Activity: Write the following numbers in expanded form.

1. 293 _____
2. 307 _____
3. 598 _____
4. 110 _____
5. 560 _____
6. 1,098 _____

Student Workbook • Number Sense   Unit 1 Addition   43

PART D
BASIC AND EXTENDED FACTS

Activity: Solve the following basic and extended facts.

1. 20 + 90 = _____   9 + 2 = _____   200 + 900 = _____
2. 300 + _____ = 1,200   120 = 30 + _____   3 + _____ = 12
3. 15 = 7 + _____   150 = _____ + 70   700 + _____ = 1,500

PART E
SOLVING IN EXPANDED AND VERTICAL FORM

Activity: Write the following equations in expanded and vertical form and solve.

EXAMPLE

```
  37   →    30 | 7
+ 42      + 40 | 2
          ─────┼───
            70 | 9
```

1. 31   →   Answer _____
  + 42

2. 537 →   Answer _____
  + 61

3. 325 →   Answer _____
  + 214

4. 437 →   Answer _____
  + 229

44 Unit 1 Addition • Lesson 20   Student Workbook • *Number Sense*

PART F
EXPANDED HORIZONTAL FORM

Activity: Write the following equations in expanded and horizontal form, rearrange, combine, and solve.

EXAMPLE

$$43 + 12 \rightarrow \underline{40 + 3} + \underline{10 + 2}$$
$$\underline{40 + 10} + \underline{3 + 2}$$
$$\underline{50} + \underline{5}$$

Answer __55__

1. $28 + 61 \rightarrow$ _____ + _____

   _____ + _____

   _____ + _____

   Answer ____

2. $37 + 22 \rightarrow$ _____ + _____

   _____ + _____

   _____ + _____

   Answer ____

3. $152 + 35 \rightarrow$ _____ + _____

   _____ + _____ + _____

   _____ + _____ + _____

   Answer ____

PART G
ESTIMATION

Activity: Round the following numbers using the front-end strategy. Find the exact answer on the calculator and compare.

EXAMPLE Estimate: 47 + 24 → 50 + 20 → 70

Exact Answer: 47 + 24 = 71

1. 11 + 59   Estimate: _____   Exact Answer: _____
2. 199 + 24   Estimate: _____   Exact Answer: _____
3. 387 + 421   Estimate: _____   Exact Answer: _____
4. 501 + 211   Estimate: _____   Exact Answer: _____

PART H
ERRORS IN ADDITION

Activity: Find the problems that contain errors and correct them.

1. $\overset{1}{3}7$
   $+ 17$
   $\overline{54}$

2. $\overset{1}{1}8$
   $+ 88$
   $\overline{116}$

3. $\overset{11}{1}85$
   $+ 98$
   $\overline{273}$

FACT FAMILIES IN SUBTRACTION
Fact-Tac-Toe

Name _____

LESSON 21

PART A
WARM-UP ACTIVITY: BASIC AND EXTENDED FACTS

Activity: Solve the following basic and extended facts. Fill in the value for the ?.

1. 7 + 2 = ____   20 + 70 = ____   700 + 200 = ____
2. ____ − 2 = 7   ____ − 20 = 70   200 = ____ − 700
3. 3 + ____ = 9   ____ + 30 = 90   30 + ____ = 90
4. 9 − 3 = ____   90 − 30 = ____   900 − 300 = ____

PART B
WRITING ADDITION/SUBTRACTION FACT FAMILIES

Activity: Write fact families for the following numbers.

1. 7, 8, and 15

   ____ + ____ = ____
   ____ + ____ = ____
   ____ − ____ = ____
   ____ − ____ = ____

2. 2, 9, and 11

   ____ + ____ = ____
   ____ + ____ = ____
   ____ − ____ = ____
   ____ − ____ = ____

3. 6, 7, and 13

   ____ + ____ = ____
   ____ + ____ = ____
   ____ − ____ = ____
   ____ − ____ = ____

4. 9, 5, and ? (Fill in the value for the ?.)

   ____ + ____ = ____
   ____ + ____ = ____
   ____ − ____ = ____
   ____ − ____ = ____

Student Workbook • Number Sense   Unit 2 Subtraction   47

PART C
EXPANDED SUBTRACTION

Activity: Write the following in expanded form and solve.

1. 69 → _____ | _____   Answer _____
    − 27   −

2. 97 → _____ | _____   Answer _____
    − 75   −

3. 79 → _____ | _____   Answer _____
    − 16   −

PART D
PLAYING "EXTENDED FACT-TAC-TOE" GAME

Activity: Use the game sheets on the following page to play the game described in the textbook.

EXTENDED FACT-TAC-TOE GAME SHEET

	Game 1	
120 − 80	80 + 40	120 − 30
150 − 70	80 + 70	90 + 30
130 − 50	50 + 80	70 + 80

	Game 2	
80 + 60	30 + 90	120 − 90
140 − 80	70 + 60	130 − 60
120 − 30	60 + 80	140 − 60

	Game 3	
130 − 40	140 − 90	140 − 50
90 + 40	130 − 90	50 + 90
160 − 90	160 − 70	70 + 90

EXPANDED SUBTRACTION
Reading Data From a Table

LESSON 22

Name _____

PART A
WARM-UP ACTIVITY: FACT FAMILIES

Activity: Solve the following fact families. Fill in the value for the ?.

1. 7 + 8 = ____
   8 + 7 = ____
   ____ − 7 = 8
   ____ − 8 = 7

2. 6 + 9 = ____
   9 + 6 = ____
   ____ − 6 = 9
   ____ − 9 = 6

3. ____ + 6 = 13
   6 + ____ = 13
   13 − 6 = ____
   13 − ____ = 6

PART B
EXPANDED SUBTRACTION

Activity: Write the following in expanded form and solve.

1. 69
   − 27 →   ____ − ____ | ____   Answer ____

2. 57
   − 36 →   ____ − ____ | ____   Answer ____

3. 96
   − 32 →   ____ − ____ | ____   Answer ____

PART C
EXTENDED FACT FAMILIES

Activity: Write extended fact families for the following numbers.

1. 70, 80, and 150

    ___ + ___ = ___

    ___ + ___ = ___

    ___ − ___ = ___

    ___ − ___ = ___

2. 200, 900, and 1,100

    ___ + ___ = ___

    ___ + ___ = ___

    ___ − ___ = ___

    ___ − ___ = ___

3. 60, 90, and 150

    ___ + ___ = ___

    ___ + ___ = ___

    ___ − ___ = ___

    ___ − ___ = ___

4. 300, 900, and ? (Fill in the missing value for the ?.)

    ___ + ___ = ___

    ___ + ___ = ___

    ___ − ___ = ___

    ___ − ___ = ___

Unit 2 Subtraction • Lesson 22   Student Workbook • *Number Sense*

MORE EXPANDED SUBTRACTION
Front-End Rounding in Subtraction

LESSON 23

Name _____

PART A
WARM-UP ACTIVITY: BASIC AND EXTENDED FACTS

Activity: Solve the following basic and extended facts. Fill in the value for the ?.

1. 15 − 9 = ____   150 − 90 = ____   1,500 − 900 = ____
2. 17 − ____ = 8   170 − 80 = ____   800 = 1,700 − ____
3. ____ − 4 = 7   70 = ____ − 40   ____ − 700 = 400
4. 11 − 5 = ____   110 − 50 = ____   1,100 − 500 = ____

PART B
EXPANDED SUBTRACTION

Activity: Write the following in expanded form and solve.

1. 428 →   Answer ____
   − 216

2. 367 →   Answer ____
   − 153

3. 589 →   Answer ____
   − 372

PART C
FRONT-END ROUNDING

Activity: Estimate the answer to the following problem using front-end rounding. Use the number lines to help you with your estimate. Then find the exact answer with your calculator and compare it to your estimate.

$$\begin{array}{r} 56 \\ -\ 17 \\ \hline \end{array}$$

We round 56 to _____.

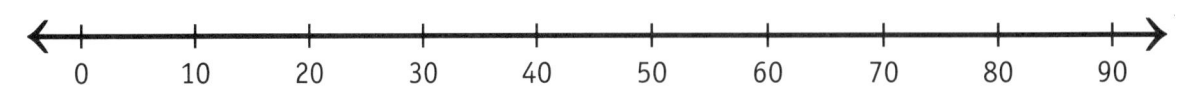

<----+----+----+----+----+----+----+----+----+---->
  0   10   20   30   40   50   60   70   80   90

We round 17 to _____.

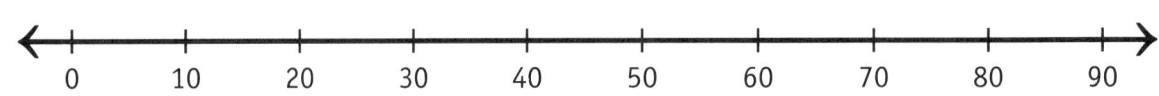

<----+----+----+----+----+----+----+----+----+---->
  0   10   20   30   40   50   60   70   80   90

The extended fact equation is: _____.

Using the calculator, the exact answer is _____.

How does your estimate compare to the exact answer? Explain why it is larger or smaller.

_____
_____
_____
_____
_____
_____
_____

REGROUPING WITH EXPANDED SUBTRACTION
Number Lines With Different Scales

LESSON 24

Name _____

PART A
WARM-UP ACTIVITY: BASIC AND EXTENDED FACT FAMILIES

Activity: Solve the following basic and extended fact families. Fill in the value for the ?.

1. 5 + 6 = ____
   6 + 5 = ____
   ____ − 5 = 6
   ____ − 6 = 5

2. 50 + 60 = ____
   60 + 50 = ____
   ____ − 50 = 60
   ____ − 60 = 50

3. 600 + 500 = ____
   500 + 600 = ____
   ____ − 500 = 600
   ____ − 600 = 500

PART B
EXPANDED SUBTRACTION WITH REGROUPING

Activity: Write the following in expanded form and solve. Show the regrouping step.

1. 64 →
    − 16

    Answer ____

2. 75 →
    − 59

    Answer ____

Student Workbook • Number Sense   Unit 2 Subtraction   53

3. $\begin{array}{r}82\\-77\end{array}$ →

Answer _____

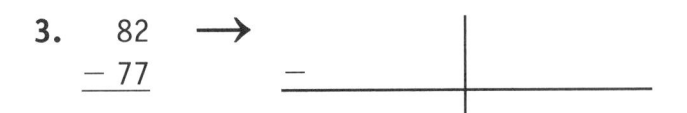

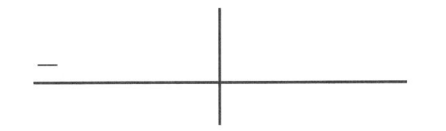

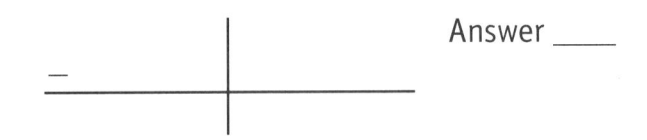

PART C
DIFFERENT SCALES ON THE NUMBER LINE

Activity 1: Create a scale for the following number lines according to the instructions given.

1. Show a number line with a scale of 10.

   0

2. Show a number line with a scale of 100.

   0

3. Show a number line with a scale of 25.

   0

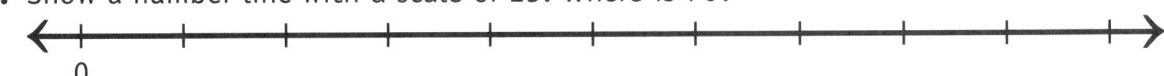

Activity 2: Create a scale for the following number lines according to the instructions given. Then mark the location of the number that is specified.

1. Show a number line with a scale of 10. Where is 25?

   0

2. Show a number line with a scale of 100. Where is 750?

   0

3. Show a number line with a scale of 25. Where is 76?

   0

4. Show a number line with a scale of 25. Where is 55?

   0

COMPARING TWO METHODS OF REGROUPING IN SUBTRACTION

Quarter Rounding Strategy

LESSON 25

Name _____

Part A
WARM-UP ACTIVITY: EXTENDED FACT FAMILIES

Activity: Solve the following extended fact families. Fill in the value for the ?.

1. 90 + 80 = ____
   80 + 90 = ____
   ____ − 90 = 80
   ____ − 80 = 90

2. 600 + 900 = ____
   900 + 600 = ____
   ____ − 60 = 90
   ____ − 90 = 60

3. ____ + 6,000 = 13,000
   6,000 + ____ = 13,000
   13,000 − 6,000 = ____
   13,000 − ____ = 6,000

PART B
EXPANDED SUBTRACTION WITH REGROUPING

Activity: Write the following in expanded form and solve.

1. 52 →
   − 18

   Answer ____

Student Workbook • *Number Sense* — Unit 2 Subtraction — 55

2. $\begin{array}{r}91\\-43\end{array}$ →

Answer ____

PART C
TRADITIONAL SUBTRACTION

Activity: Use the traditional method of subtraction to solve the following problems.

1. $\begin{array}{r}452\\-127\end{array}$

2. $\begin{array}{r}564\\-83\end{array}$

PART D
WORKING WITH DIFFERENT SCALES ON NUMBER LINES

Activity: Create a scale of 25 for each of the number lines. Mark the location of the given number and find the nearest quarter.

1. Where is 105? What is the nearest quarter?

   ←+—+—+—+—+—+—+—+—+—+—+→
   0

2. Where is 160? What is the nearest quarter?

   ←+—+—+—+—+—+—+—+—+—+—+→
   0

3. Where is 195? What is the nearest quarter?

   ←+—+—+—+—+—+—+—+—+—+—+→
   0

4. Where is 29? What is the nearest quarter?

   ←+—+—+—+—+—+—+—+—+—+—+→
   0

5. Where is 109? What is the nearest quarter?

   ←+—+—+—+—+—+—+—+—+—+—+→
   0

MORE PRACTICE REGROUPING IN SUBTRACTION
Comparing Rounding Strategies

LESSON 26

Name _____

PART A
WARM-UP ACTIVITY: EXTENDED FACT FAMILIES

Activity: Solve the following extended fact families. Fill in the value of X.

1. $70 + 90 =$ ____
   $90 + 70 =$ ____
   ____ $- 90 = 70$
   ____ $- 70 = 90$

2. $600 + 800 =$ ____
   $800 + 600 =$ ____
   ____ $- 60 = 80$
   ____ $- 80 = 60$

3. ____ $+ 5{,}000 = 12{,}000$
   $5{,}000 +$ ____ $= 12{,}000$
   $12{,}000 - 5{,}000 =$ ____
   $12{,}000 -$ ____ $= 5{,}000$

PART B
EXPANDED SUBTRACTION WITH REGROUPING

Activity: Write the following in expanded form and solve.

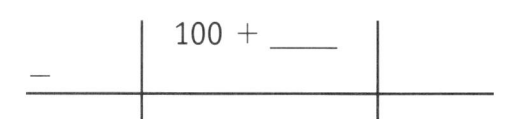

1. 352 →
   $- 181$

   ____ $+ 100$

   $100 +$ ____

   Answer ____

2. 915 →
   − 783

   _____ − _____ | _____ | _____

   _____ + 100 | _____
   _____ − _____ | _____

   _____ − _____ | 100 + _____ | _____

   _____ − _____ | _____ | _____   Answer _____

PART C
COMPARING METHODS FOR ROUNDING NUMBERS

Activity: In the following problems, decide whether to use front-end rounding or rounding to nearest quarter and come up with an estimated answer for the problem. Then find the actual answer with your calculator.

Problem	Strategy	Estimate	Actual Answer
1. 79 − 19			
2. 297 − 98			
3. 382 − 157			
4. 591 − 319			
5. 972 − 727			

Unit 2 Subtraction • Lesson 26   Student Workbook • *Number Sense*

COMPARING REGROUPING METHODS
More Practice Reading Data From Tables

LESSON 27

Name _____

PART A
WARM-UP ACTIVITY:
ADDITION AND SUBTRACTION QUARTER FACTS

Activity: Solve the following addition and subtraction quarter facts. Fill in the value of X.

1. $75 - 25 =$ _____
2. $25 + 50 =$ _____
3. $225 - 25 =$ _____
4. $100 -$ _____ $= 50$
5. _____ $- 25 = 75$
6. $150 +$ _____ $= 200$
7. $75 +$ _____ $= 100$
8. $100 + 25 =$ _____
9. $350 -$ _____ $= 300$

PART B
COMPARING TRADITIONAL AND EXPANDED SUBTRACTION

Activity 1: Use the traditional method of subtraction to solve the following problem.

$$\begin{array}{r} 527 \\ -\ 235 \\ \hline \end{array}$$

Activity 2: Use expanded form to solve the problem that you solved in Activity 1.

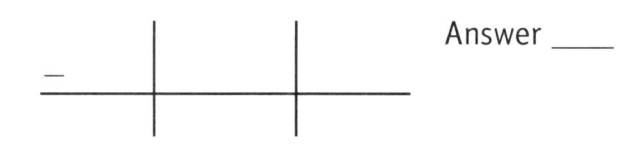

$$\begin{array}{r} 527 \\ -235 \end{array} \rightarrow$$

___ + 100

100 + ___

Answer ___

Explain what is different about the expanded subtraction and traditional subtraction methods.

_____
_____
_____
_____
_____
_____

REVIEW OF LESSONS 21 TO 27

Lesson 28

Name _____

PART A
WARM-UP ACTIVITY: BASIC AND EXTENDED FACT FAMILIES AND QUARTER FACTS

Activity: Solve the following basic and extended fact families and quarter facts. Fill in the value of X.

1. $7 + 9 =$ ____
   $9 + 7 =$ ____
   ____ $- 7 = 9$
   ____ $- 9 = 7$

2. $70 + 90 =$ ____
   $90 + 70 =$ ____
   ____ $- 70 = 90$
   ____ $- 90 = 70$

3. $25 + 75 =$ ____
   $75 + 25 =$ ____
   ____ $- 25 = 75$
   ____ $- 75 = 25$

PART B
EXPANDED SUBTRACTION

Activity: Write the following in expanded form and solve.

1. 28
   -16 →   ____ | ____   Answer ____

2. 367
   -53 →   ____ | ____ | ____   Answer ____

3. 489
   -353 →   ____ | ____ | ____   Answer ____

Student Workbook • Number Sense   Unit 2 Subtraction   63

PART C
ROUNDING STRATEGIES

Activity: Decide which rounding strategy to use for these problems. Come up with an estimated answer for each problem. Then find the exact answer with your calculator.

Problem	Rounding Strategy	Estimate	Calculator Answer
1. 572 − 326			
2. 399 − 201			
3. 454 − 227			
4. 891 − 699			

PART D
REGROUPING IN SUBTRACTION

Activity: Solve using any method.

1. 52
   − 27

2. 764
   − 83

3. 852
   − 427

4. 964
   − 291

IDENTIFYING COMMON SUBTRACTION ERRORS

The Blockheads Concert Tour: Florida

Name _____

LESSON 29

PART A
WARM-UP ACTIVITY: BASIC AND EXTENDED FACT FAMILIES

Activity: Solve the following basic and extended fact families. Fill in the value of Y.

1. $4 + 8 =$ ____ $8 + 4 =$ ____   ____ $- 4 = 8$ ____ $- 8 = 4$
2. $40 + 80 =$ ____ $80 + 40 =$ ____   ____ $- 40 = 80$ ____ $- 80 = 40$
3. $6 + 7 =$ ____ $7 + 6 =$ ____   ____ $- 7 = 6$ ____ $- 6 = 7$
4. $60 + 70 =$ ____ $70 + 60 =$ ____   ____ $- 60 = 70$ ____ $- 70 = 60$

PART B
COMMON SUBTRACTION ERRORS

Activity 1: Find the errors in the following two problems. Rewrite the problems in expanded form to help you understand the errors.

1. 351
   − 117
   ─────
    246   Answer ____

2. 404
   − 27
   ─────
    427   Answer ____

Activity 2: Describe the errors that the student made in these two problems.

MORE ON COMMON ERRORS
Double-Check the Bill

LESSON 30

Name _____

PART A
WARM-UP ACTIVITY: BASIC AND EXTENDED FACT FAMILIES

Activity: Solve the following basic and extended fact families. Fill in the value of Y.

1. $4 + 9 =$ ____ $9 + 4 =$ ____   ____ $- 4 = 9$ ____ $- 9 = 4$
2. $40 + 90 =$ ____ $90 + 40 =$ ____   ____ $- 40 = 90$ ____ $- 90 = 40$
3. $6 + 5 =$ ____ $5 + 6 =$ ____   ____ $- 5 = 6$ ____ $- 6 = 5$
4. $60 + 50 =$ ____ $50 + 60 =$ ____   ____ $- 50 = 60$ ____ $- 60 = 50$

PART B
COMMON ERRORS IN SUBTRACTION

Activity 1: Find the errors in the following two problems. Rewrite the problems in expanded form to help you understand the errors.

1. 896 →   Answer ____
   − 248
   ─────
    658

2. ⁵⁹6̸0̸1 →   Answer ____
   − 240
   ─────
    351

Student Workbook • *Number Sense*   Unit 2 Subtraction   **67**

Activity 2: Describe the errors that the student made in these two problems.

THE TROUBLE WITH ZEROS IN SUBTRACTION

LESSON 31

Name _____

Problem Solving: What's the Problem Asking For?

PART A
WARM-UP ACTIVITY:
WRITING FACT FAMILIES FOR QUARTER FACTS

Activity: Solve the following quarter fact families.

1. 25, 75, and 100

   ____ + ____ = ____

   ____ + ____ = ____

   ____ − ____ = ____

   ____ − ____ = ____

2. 125, 50, and 175

   ____ + ____ = ____

   ____ + ____ = ____

   ____ − ____ = ____

   ____ − ____ = ____

PART B
SUBTRACTION ESTIMATIONS

Activity: Solve the following problems using any method. Use estimation and your calculator to check your answer.

1. 302
   − 143

   Estimation:

   Calculator Check:

2. 8,001
   − 6,233

   Estimation:

   Calculator Check:

Student Workbook • Number Sense — Unit 2 Subtraction

3. 6,002
   − 1,043

Estimation:

Calculator Check:

4. 9,900
   − 999

Estimation:

Calculator Check:

5. 5,100
   − 88

Estimation:

Calculator Check:

WHY WE LEARN TO ESTIMATE

The Blockheads Concert Tour: On to Texas

LESSON 32

Name _____

PART A
WARM-UP ACTIVITY: BASIC AND EXTENDED FACT FAMILIES IN OPEN SENTENCES

Activity: Solve the following basic and extended facts. Fill in the value of Z.

1. 7 + 8 = ____   8 + 7 = ____   ____ − 7 = 8   ____ − 8 = 7
2. 70 + 80 = ____   80 + 70 = ____   ____ − 70 = 80   ____ − 80 = 70
3. 2 + 9 = ____   9 + 2 = ____   ____ − 2 = 9   ____ − 9 = 2
4. 200 + 900 = ____   900 + 200 = ____   ____ − 200 = 900   ____ − 900 = 200

PART B
ESTIMATION AND NUMBER LINES

Activity: Estimate the answer to the following problems using the number lines to help you with your estimate. Then compute the exact answer using your calculator.

1. 3,056
   − 1,917

   We round 3,056 to ____.

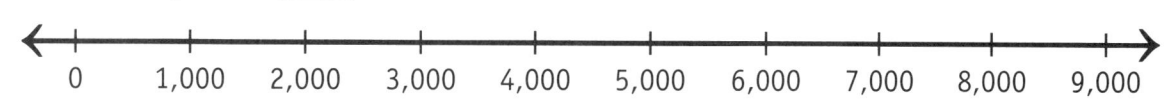

   $\longleftarrow$ 0 1,000 2,000 3,000 4,000 5,000 6,000 7,000 8,000 9,000 $\longrightarrow$

   We round 1,917 to ____.

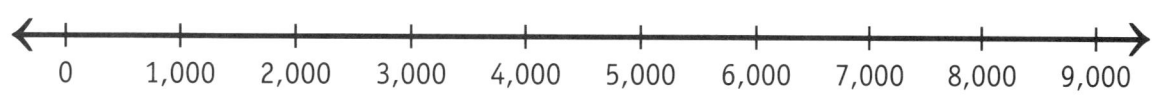

   $\longleftarrow$ 0 1,000 2,000 3,000 4,000 5,000 6,000 7,000 8,000 9,000 $\longrightarrow$

   The extended fact equation is _____.

   Using the calculator, the exact answer is ____.

Student Workbook • Number Sense   Unit 2 Subtraction   71

2. 4,001
   − 3,078

   We round 4,001 to ____.

   ←|—+—+—+—+—+—+—+—+—+—→
   0 1,000 2,000 3,000 4,000 5,000 6,000 7,000 8,000 9,000

   We round 3,078 to ____.

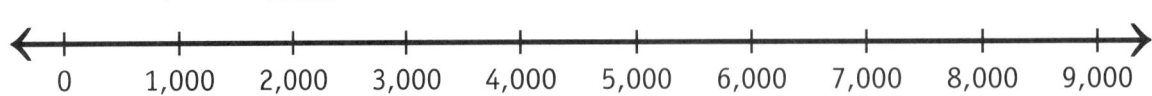

   ←|—+—+—+—+—+—+—+—+—+—→
   0 1,000 2,000 3,000 4,000 5,000 6,000 7,000 8,000 9,000

   The extended fact equation is _____.

   Using the calculator, the exact answer is ____.

3. 10,056
   − 4,895

   We round 10,056 to ____.

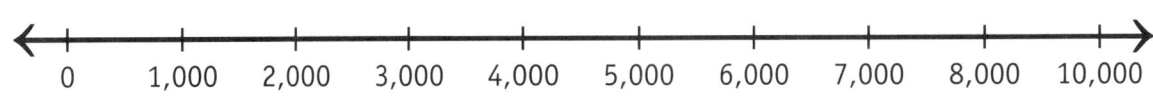

   ←|—+—+—+—+—+—+—+—+—+—→
   0 1,000 2,000 3,000 4,000 5,000 6,000 7,000 8,000 10,000

   We round 4,895 to ____.

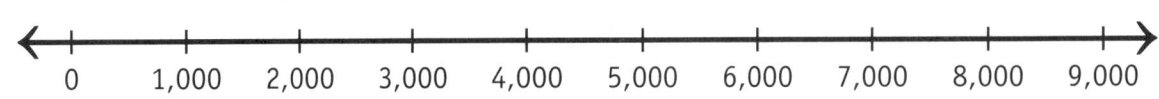

   ←|—+—+—+—+—+—+—+—+—+—→
   0 1,000 2,000 3,000 4,000 5,000 6,000 7,000 8,000 9,000

   The extended fact equation is _____.

   Using the calculator, the exact answer is ____.

USING GOOD NUMBER SENSE

The Bottom Line: Did We Make Any Money?

Lesson 33

Name _____

PART A
WARM-UP ACTIVITY: NUMBERS WITH UNIQUE DIFFERENCES

Activity: Solve the subtraction facts. Fill in the value of Z.

1. 10 − 9 = ____   100 − 99 = ____   1,000 − 999 = ____
2. 10 − 5 = ____   100 − 5 = ____   1,000 − 5 = ____
3. 10 − 1 = ____   100 − 1 = ____   1,000 − 1 = ____
4. 10 − 5 = ____   100 − 95 = ____   1,000 − 995 = ____

PART B
USING GOOD NUMBER SENSE TO SOLVE PROBLEMS

Activity: Use number lines to help you solve the following problems without using regrouping.

1. 500 − 299

2. 500 − 302

3. 700 − 495

4. 700 − 505

Student Workbook • Number Sense — Unit 2 Subtraction

WRITING ABOUT WHAT YOU HAVE LEARNED
How Much Does a Movie Make?

LESSON 34

Name _____

PART A
WARM-UP ACTIVITY: FACT FAMILIES WITH QUARTER FACTS

Activity: Solve the following quarter fact families. Replace the Z with its value.

1. 25, 75, and Z

   ___ + ___ = ___

   ___ + ___ = ___

   ___ − ___ = ___

   ___ − ___ = ___

2. 250, Z, and 750

   ___ + ___ = ___

   ___ + ___ = ___

   ___ − ___ = ___

   ___ − ___ = ___

PART B
CHOOSING THE BEST METHOD FOR SOLVING PROBLEMS

Activity: Choose the best method for solving the problems below. The choices are (a) Estimation and Calculator, or (b) Mental Math. Then solve using the method selected.

1. 4,905
   − 1,896

   Method: (circle one)
   (a) Estimation and Calculator
   (b) Mental Math

2. 500
   − 198

   Method: (circle one)
   (a) Estimation and Calculator
   (b) Mental Math

Student Workbook • *Number Sense* Unit 2 Subtraction

3. 257
   − 145

   Method: (circle one)
   (a) Estimation and Calculator
   (b) Mental Math

4. 1,219
   − 993

   Method: (circle one)
   (a) Estimation and Calculator
   (b) Mental Math

5. 5,001
   − 1,999

   Method: (circle one)
   (a) Estimation and Calculator
   (b) Mental Math

UNIT 2 REVIEW

Name _____

LESSON 35

PART A
WARM-UP ACTIVITY:
BASIC AND EXTENDED QUARTER FACT FAMILIES

Activity: Solve the following basic and extended quarter fact families. Replace the Z with its value.

1. 25 + 50 = __Z__
    50 + 25 = __Z__
    __Z__ − 25 = 50
    __Z__ − 50 = 25

2. 250 + 500 = __Z__
    500 + 250 = __Z__
    __Z__ − 250 = 500
    __Z__ − 500 = 250

3. 25 + __Z__ = 125
    __Z__ + 25 = 125
    125 − 25 = __Z__
    125 − __Z__ = 25

PART B
EXPANDED SUBTRACTION

Activity: Write the following in expanded form and solve.

1. 67
    − 32   →   __ − __ | __   Answer ____

2. 856
    − 45   →   __ − __ | __ | __   Answer ____

3. 776
    − 223   →   __ − __ | __ | __   Answer ____

Student Workbook • *Number Sense*   Unit 2 Subtraction   **77**

PART C
ROUNDING STRATEGIES

Activity: In the following problems, decide whether to use front-end rounding or rounding to nearest quarter and come up with an estimated answer for the problem. Then find the actual answer with your calculator.

Problem	Rounding Strategy	Estimate	Calculator Answer
1. 251 − 179			
2. 401 − 299			
3. 327 − 199			
4. 751 − 524			

PART D
REGROUPING IN SUBTRACTION

Activity: Solve using any method.

1. 86 − 27

2. 464 − 93

3. 819 − 427

4. 982 − 376

PART E
FINDING ERRORS IN SUBTRACTION

Activity: Some of the following problems contain errors. Find the errors and fix them.

1. 42
   − 27
   ────
    15

2. 464
   − 93
   ────
    431

3. 819
   − 427
   ────
    392

4. 982
   − 376
   ────
    506

PART F
CHOOSING THE BEST METHOD

Activity: Choose the best method to solve the following problems. The choices are:

(a) Traditional Subtraction

(b) Estimation and Calculator

(c) Good Number Sense and Number Line

Then Solve.

1. 4,905
   − 1,896   Method:

2. 600
   − 395   Method:

3. 874
   − 526   Method:

Student Workbook • *Number Sense* Unit 2 Subtraction • Lesson 35 **79**

BASIC AND EXTENDED FACTS
Commuting in Multiplication

LESSON 36

Name _____

PART A
WARM-UP ACTIVITY: BASIC MULTIPLICATION FACTS AND COMMUTATION

Activity: Solve the following basic multiplication facts. Notice the turn-around facts.

1. $3 \times 9 =$ ____
2. $8 \times 9 =$ ____
3. $6 \times$ ____ $= 42$
4. $9 \times 3 =$ ____
5. $3 \times$ ____ $= 27$
6. ____ $\times 4 = 36$
7. $9 \times 8 =$ ____
8. $7 \times$ ____ $= 42$
9. ____ $\times 9 = 36$

PART B
NUMBER LINES AND BASIC FACTS IN MULTIPLICATION

Activity: Make units and use them to show the following multiplication problems on the number line.

$2 \times 3 = ?$

EXAMPLE

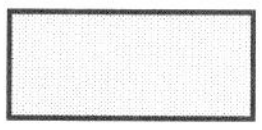

☐ This is a unit of 2 and we repeat it 3 times.

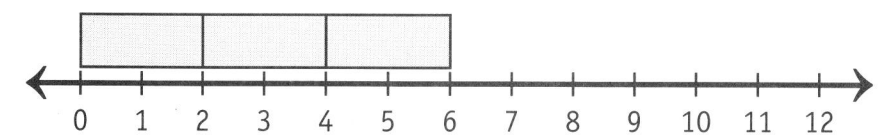

1. $3 \times 3 = ?$

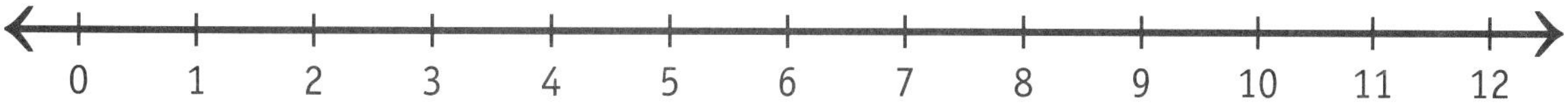

2. $5 \times 2 = ?$

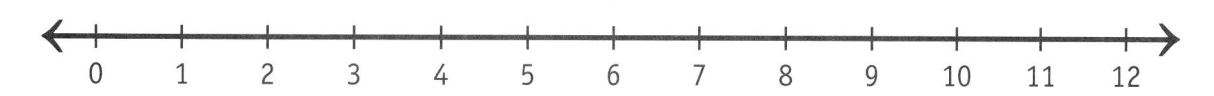

3. $2 \times 4 = ?$

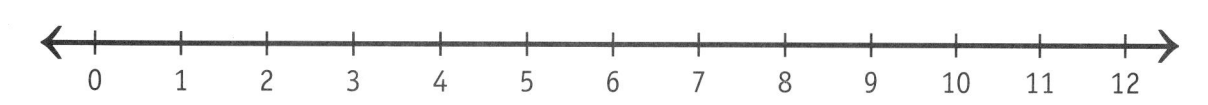

Student Workbook • *Number Sense* Unit 3 Multiplication 81

PART C
FACTS AND EXTENDED FACTS

Activity 1: Solve the following basic and extended multiplication facts.

1. 2 × 9 = ____   2 × 90 = ____   2 × 900 = ____
2. 3 × 9 = ____   3 × 90 = ____   3 × 900 = ____
3. 7 × 8 = ____   7 × 80 = ____   7 × 800 = ____
4. 4 × 6 = ____   4 × 60 = ____   4 × 600 = ____
5. 5 × 8 = ____   5 × 80 = ____   5 × 800 = ____

Activity 2: Fill in the missing value.

1. 7 × ____ = 56   7 × ____ = 560   7 × ____ = 5,600
2. 4 × ____ = 32   4 × ____ = 320   4 × ____ = 3,200
3. ____ × 9 = 81   ____ × 90 = 810   ____ × 900 = 8,100

Activity 3: Show the following extended multiplication facts on the number lines given.

EXAMPLE   20 × 3 = ? → We are going to move 20 spaces, 3 times. Our scale is counting by 10s.

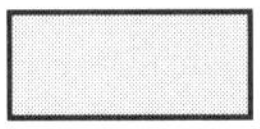

☐ This is a unit of 20 and we repeat it 3 times.

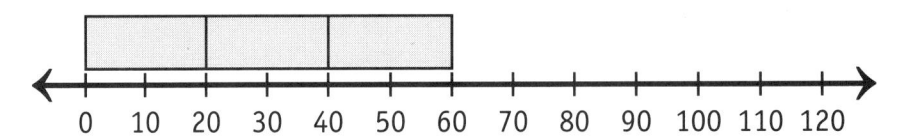

←—+—+—+—+—+—+—+—+—+—+—+—+—→
0 10 20 30 40 50 60 70 80 90 100 110 120

1. 30 × 2 = ?

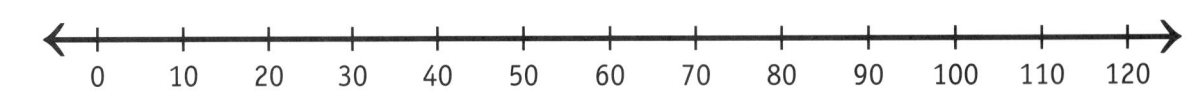

←—+—+—+—+—+—+—+—+—+—+—+—+—→
0 10 20 30 40 50 60 70 80 90 100 110 120

2. 40 × 3 = ?

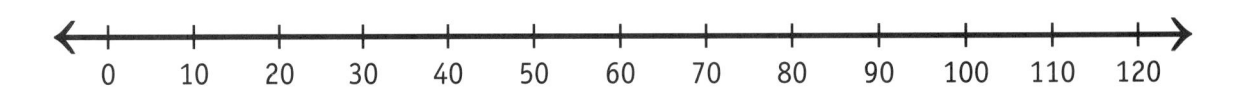

←—+—+—+—+—+—+—+—+—+—+—+—+—→
0 10 20 30 40 50 60 70 80 90 100 110 120

FINDING THE 10 IN NUMBERS
Estimating Distances

LESSON 37

Name _____

PART A
WARM-UP ACTIVITY: BASIC AND EXTENDED MULTIPLICATION FACTS

Activity: Solve the following basic and extended multiplication facts.

1. $7 \times 8 =$ ____
   $8 \times 7 =$ ____
   $7 \times 80 =$ ____
   $7 \times 800 =$ ____

2. $9 \times 3 =$ ____
   $3 \times 9 =$ ____
   $3 \times 90 =$ ____
   $3 \times 900 =$ ____

3. $4 \times 6 =$ ____
   $6 \times 4 =$ ____
   $6 \times 40 =$ ____
   $6 \times 400 =$ ____

PART B
FINDING THE 10

Activity 1: Rewrite the following as "a number times 10."

Starting Number	? × 10
50	5 × 10
60	
70	
80	
90	

Student Workbook • *Number Sense* — Unit 3 Multiplication — 83

POWERS OF 10
Making a Measuring Device

LESSON 38

Name _____

PART A
WARM-UP ACTIVITY: BASIC AND EXTENDED MULTIPLICATION FACTS

Activity: Solve the following basic and extended facts.

1. $9 \times 8 =$ ____
   $8 \times 9 =$ ____
   $9 \times 80 =$ ____
   $9 \times 800 =$ ____

2. $7 \times 3 =$ ____
   $3 \times 7 =$ ____
   $3 \times 70 =$ ____
   $3 \times 700 =$ ____

3. $5 \times 6 =$ ____
   $6 \times 5 =$ ____
   $6 \times 50 =$ ____
   $6 \times 500 =$ ____

PART B
PULLING OUT POWERS OF 10

Activity 1: Rewrite the following numbers by pulling out the largest power of 10.

Starting Number	? × Largest Power of 10
5,000	5 × 1,000
300	
6,000	
700	
40	
8,000	

Student Workbook • Number Sense — Unit 3 Multiplication — 85

Activity 2: Fill in the missing values in the table.

Number	? × 10	? × 100	? × 1,000
70	7 × 10		
80			
900	90 × 10	9 × 100	
400			
500			
6,000	600 × 10	60 × 100	6 × 1,000
1,000			
2,000			
3,000			

EXPANDED MULTIPLICATION
Making a Ruler of 1s and 10s

LESSON 39

Name _____

PART A
WARM-UP ACTIVITY: BASIC FACTS, EXTENDED FACTS, AND PULLING OUT THE 10

Activity 1: Solve the following basic and extended facts.

1. $7 \times 6 = $ ____
   $6 \times 7 = $ ____
   $7 \times 60 = $ ____
   $7 \times 600 = $ ____

2. $4 \times 5 = $ ____
   $5 \times 4 = $ ____
   $5 \times 40 = $ ____
   $5 \times 400 = $ ____

3. $7 \times 3 = $ ____
   $3 \times 7 = $ ____
   $7 \times 30 = $ ____
   $7 \times 300 = $ ____

Activity 2: Rewrite the following numbers as a number times 10. The first answer is given.

Number	? × 10
70	7 × 10
80	
90	
50	
30	
40	

Student Workbook • Number Sense · Unit 3 Multiplication · 87

Part B
EXPANDED MULTIPLICATION

Activity 1: Write the following in expanded form and solve. Look at the example carefully.

EXAMPLE

```
    95    →     90 | 5
   × 3         ×  | 3
                  15
              +  270
                 285
```

1. 86
   × 3

   × ____ | ____

2. 47
   × 6

   × ____ | ____

3. 34
   × 5

   × ____ | ____

4. 78
   × 5

   × ____ | ____

THE METRIC SYSTEM
Measuring With a Metric Ruler

Lesson 40

Name _____

PART A
WARM-UP ACTIVITY: BASIC AND EXTENDED FACTS AND FINDING THE 1,000s, 100s, AND 10s

Activity 1: Solve the following basic and extended facts.

1. 4 × ____ = 24
2. 4 × 50 = ____
3. 5 × 9 = ____
4. 5 × 90 = ____
5. ____ × 4 = 12
6. 3 × 30 = ____
7. 3 × 300 = ____
8. ____ × 6 = 42
9. 6 × 9 = ____
10. 6 × 90 = ____
11. ____ × 5 = 35
12. ____ × 5 = 350
13. 8 × 8 = ____
14. 8 × 80 = ____
15. 4 × ____ = 32
16. 4 × 40 = ____

Activity 2: Fill in the missing expressions in the table.

Number	? × 10	? × 100	? × 1,000
4,000	400 × 10	40 × 100	4 × 1,000
5,000			
6,000			

Student Workbook • Number Sense Unit 3 Multiplication 89

PART B

Activity: Write the following problems in expanded form and solve.

1. 37
   × 5

   × _____|_____

2. 82
   × 7

   × _____|_____

3. 65
   × 3

   × _____|_____

PART C
THE METRIC SYSTEM: PRACTICE USING RULERS

Activity: Write the length of these lines in the spaces provided. Be sure to measure with the metric unit specified—*millimeter (mm), centimeter (cm), or decimeter (dm)*. For example, Line A should be measured using centimeters. Line B should be measured using millimeters.

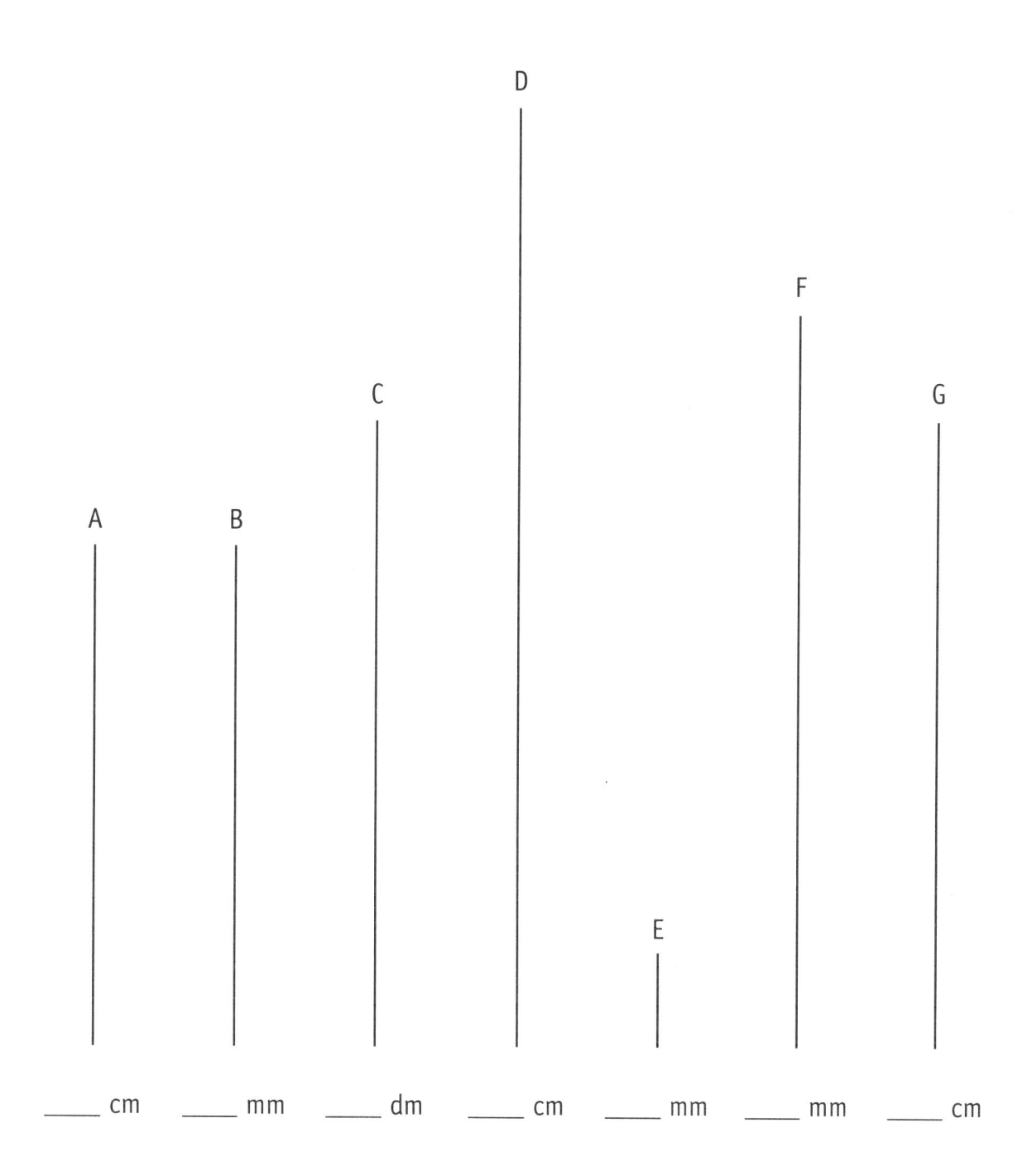

____ cm   ____ mm   ____ dm   ____ cm   ____ mm   ____ mm   ____ cm

Student Workbook • *Number Sense*   Unit 3 Multiplication • Lesson 40   **91**

REVIEW OF LESSONS 36 TO 40

Name _____

Lesson 41

PART A
WARM-UP ACTIVITY: BASIC AND EXTENDED FACTS AND COMMUTING IN MULTIPLICATION

Activity 1: Solve the following basic and extended facts. Notice the turn-around facts.

1. $4 \times 5 = $ _____
2. $6 \times 4 = $ _____
3. $3 \times 9 = $ _____
4. $7 \times 8 = $ _____

5. $4 \times 50 = $ _____
6. $6 \times 40 = $ _____
7. $3 \times 90 = $ _____
8. $8 \times 7 = $ _____

9. $5 \times 4 = $ _____
10. $6 \times 400 = $ _____
11. $3 \times 900 = $ _____
12. $8 \times 70 = $ _____

13. $5 \times 40 = $ _____
14. $4 \times 6 = $ _____
15. $9 \times 3 = $ _____
16. $8 \times 700 = $ _____

PART B
FINDING THE 1,000s, 100s, AND 10s

Activity: Fill in the missing expressions in the chart. Note that some numbers cannot be rewritten using all of the columns.

Number	? × 10	? × 100	? × 1,000
400	40 × 10	4 × 100	
500			
8,000			
60			
6,000			
90			
500			

Student Workbook • Number Sense   Unit 3 Multiplication   93

PART C
MULTIPLICATION IN EXPANDED FORM

Activity: Write the following problems in expanded form and solve.

1. 48
   × 6

   × _____ | _____

2. 95
   × 4

   × _____ | _____

3. 93
   × 4

   × _____ | _____

4. 88
   × 3

   × _____ | _____

COMPARING METHODS OF REGROUPING
The World of Design

LESSON 42

Name _____

PART A
WARM-UP ACTIVITY: BASIC FACTS AND EXTENDED FACTS

Activity: Solve the following basic and extended facts. Write your answer over the Z.

1. 4 × ____ = 20
2. ____ × 4 = 24
3. ____ × 9 = 81
4. 7 × ____ = 56
5. 4 × 50 = ____
6. ____ × 6 = 24
7. 3 × 90 = ____
8. 8 × 7 = ____
9. 5 × ____ = 35
10. 6 × ____ = 240
11. 3 × 900 = ____
12. 8 × ____ = 560

PART B
TRADITIONAL AND EXPANDED MULTIPLICATION

Activity 1: Solve the following multiplication problems using the traditional method.

1. 47
   × 8

2. 35
   × 4

Activity 2: Solve the same two problems using expanded multiplication.

1. 47
   × 8

2. 35
   × 4

Student Workbook • Number Sense   Unit 3 Multiplication   95

Activity 3: Compare the two methods. Write at least two things that are the same about the methods and at least two things that are different.

1. How are the methods the same?

   _____

   _____

   _____

   _____

2. How are the methods different?

   _____

   _____

   _____

   _____

ROUNDING STRATEGIES IN MULTIPLICATION
Making a Logo

LESSON 43

Name _____

PART A
WARM-UP ACTIVITY: POWERS OF 10

Activity: The following problems all involve powers of 10. Find the value of y in each problem. Remember that we use variables (letters) to stand for missing values. Find the missing value for y. Note that the * is another symbol for multiplication.

1. 50 = 5 * y
2. 500 = 5 * y
3. 500 = 50 * y
4. 700 = 7 * y

5. 700 = 70 * y
6. 70 = 7 * y
7. 400 = 4 * y
8. 400 = 40 * y

9. 40 = 4 * y
10. 80 = 8 * y
11. 800 = 80 * y
12. 800 = 8 * y

PART B
ESTIMATION IN MULTIPLICATION

Activity: Use rounding to make extended facts for the following multiplication problems and solve the extended facts. Show your estimation on a number line. Then find the actual answer on your calculator and compare.

376   376 is about 400   400 × 5 = 2,000   **EXAMPLE**
× 5   My estimation is 2,000.

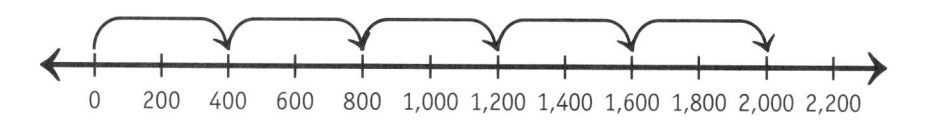

Actual Answer: 1,880

1. 29 → 29 is about ____ → ____ × 4 = ____
   × 4

   Estimation ____
   Actual Answer ____

   ←+—+—+—+—+—+—+—+—+—+—+—+—+→
    0 10 20 30 40 50 60 70 80 90 100 110 120

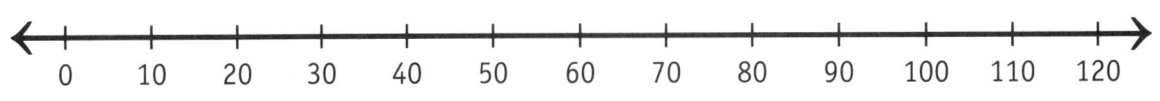

2. 219 → 219 is about ____ → ____ × 5 = ____
   × 5

   Estimation ____
   Actual Answer ____

   ←+—+—+—+—+—+—+—+—+—+—+—+—+→
    0 100 200 300 400 500 600 700 800 900 1,000 1,100 1,200

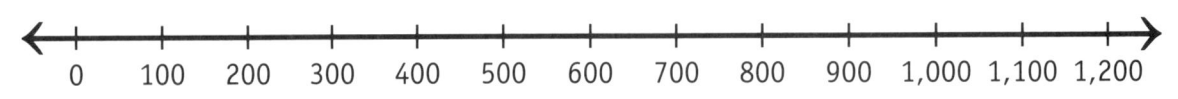

3. 792 → 792 is about ____ → ____ × 3 = ____
   × 3

   Estimation ____
   Actual Answer ____

   ←+—+—+—+—+—+—+—+—+—+—+—+—+→
    0 200 400 600 800 1,000 1,200 1,400 1,600 1,800 2,000 2,200 2,400

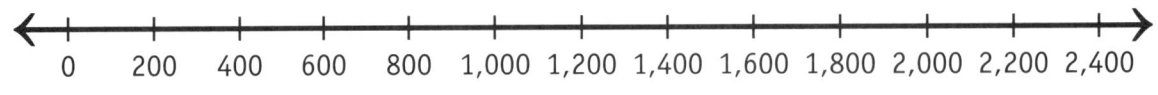

4. 209 → 209 is about ____ → ____ × 6 = ____
   × 6

   Estimation ____
   Actual Answer ____

   ←+—+—+—+—+—+—+—+—+—+—+—+—+→
    0 100 200 300 400 500 600 700 800 900 1,000 1,100 1,200

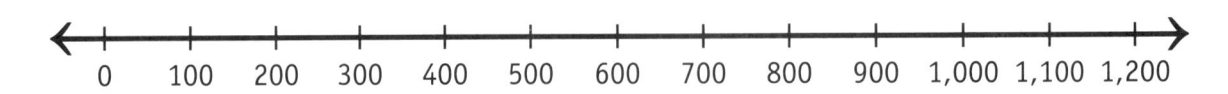

PART C
DRAWING THE TRIANGLE PART OF THE LOGO

Activity: Look at the design in the textbook. Draw the design in the space below. Be sure to do the following:

1. Write **Sketch of the Triangle Part of the Logo** at the top of the page.

2. Make sure the triangles match the drawing. In the space below, we have started the sides of the five triangles that are at the bottom. Make sure that two long sides of each triangle are 5 cm long. Also, the two long sides of the top triangle should be 5 cm long.

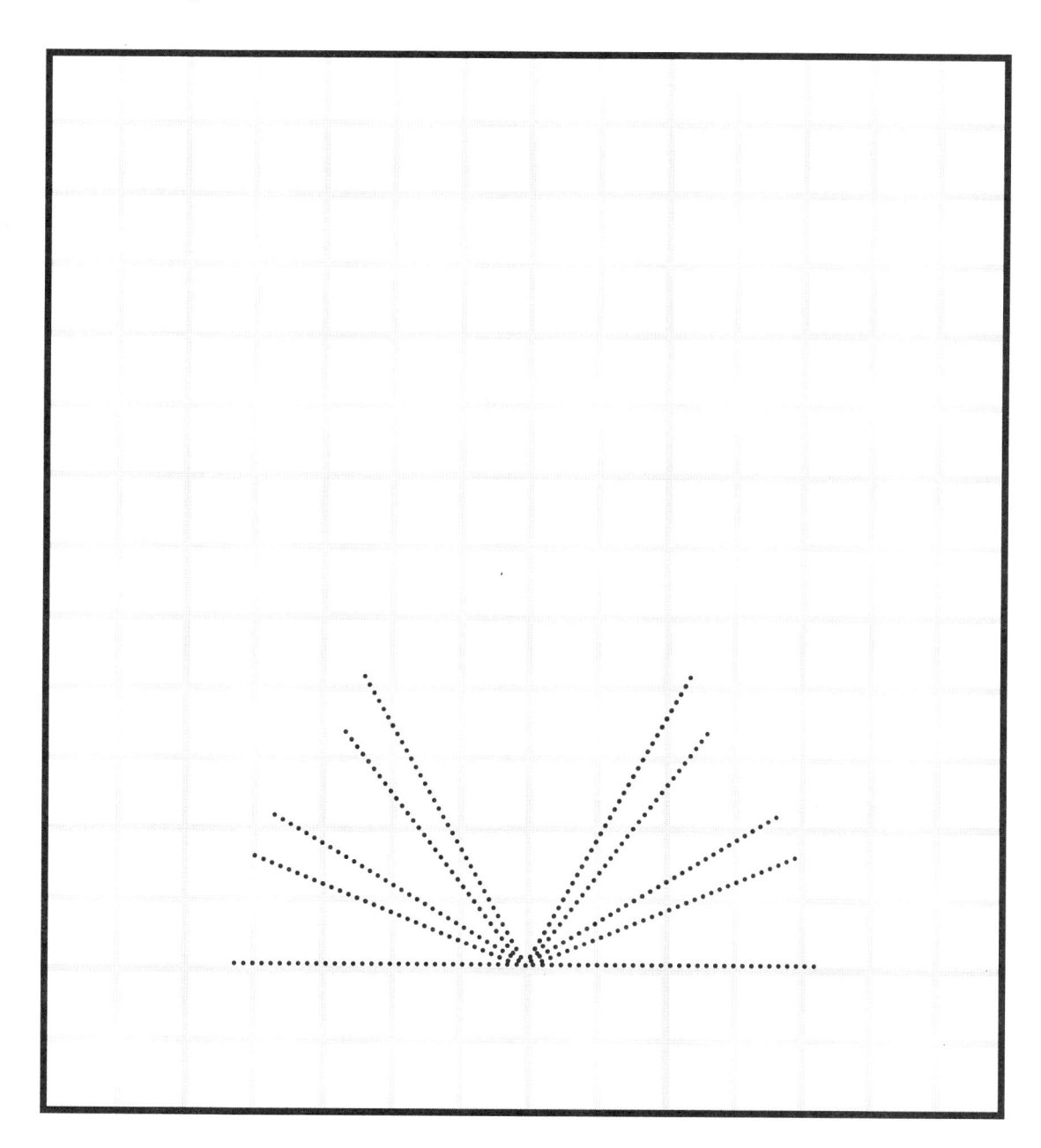

Student Workbook • *Number Sense* — Unit 3 Multiplication • Lesson 43

WHY WE MAKE MISTAKES IN MULTIPLICATION
Estimation Tasks

LESSON 44

Name _____

PART A
WARM-UP ACTIVITY: BASIC AND EXTENDED MULTIPLICATION FACTS

Activity: Solve the following basic and extended multiplication facts.

1. $4 \times 3 =$ ____
2. $4 \times 30 =$ ____
3. $4 \times 300 =$ ____
4. $7 \times 8 =$ ____

5. $8 \times 7 =$ ____
6. $8 \times 70 =$ ____
7. $8 \times 700 =$ ____
8. $7 \times 80 =$ ____

9. $3 \times 4 =$ ____
10. $3 \times 40 =$ ____
11. $3 \times 400 =$ ____
12. $6 \times 5 =$ ____

13. $6 \times 50 =$ ____
14. $6 \times 500 =$ ____
15. $5 \times 6 =$ ____
16. $5 \times 60 =$ ____

PART B
SIMPLE MULTIPLICATION ERRORS

Activity: Which of these problems contain errors? Once you find the errors, write about them in the spaces provided.

1. $\overset{3}{39} \\ \underline{\times\ 7} \\ 246$

2. $\overset{2}{231} \\ \underline{\times\ 9} \\ 2{,}079$

3. $\overset{5}{89} \\ \underline{\times\ 6} \\ 484$

_____
_____
_____
_____
_____
_____

Student Workbook • *Number Sense*　　　Unit 3 Multiplication　　**101**

PART C
ERRORS ON A CALCULATOR

Activity: A student used a calculator to work the following problems, but there are still mistakes. Which of these answers are incorrect? Use your calculator to check the student's work and figure out what the mistakes are. In the spaces provided, describe the error(s).

1.   275
      × 43
     11,825

2.   809
      × 19
     73,619

3.   80
      × 6
     4,800

PART D
ESTIMATING HEIGHT

Activity: For the following problems, circle the best answer and write one or two sentences that explain why you chose the answer you did.

1. How tall is the goal?

   a. 20 centimeters

   b. 2 meters

   c. 15 millimeters

   Tell why:

   _____

   _____

   _____

   _____

2. How tall is the elephant?

   a. 4 meters

   b. 20 meters

   c. 150 centimeters

   Tell why:

   _____

   _____

   _____

   _____

   _____

   _____

   _____

Student Workbook • *Number Sense* Unit 3 Multiplication • Lesson 44

3. How tall is the ladder?

   a. 10 meters

   b. 40 meters

   c. 65 meters

   Tell why:

   _____

   _____

LAYOUT AND DESIGN
Making a Web Page

Name _____

LESSON 45

PART A
WARM-UP ACTIVITY: POWERS OF 10

Activity: Fill in the missing values in the following table.

Note: Put an X in the squares where you cannot fill in a powers of 10 number.

Number	? × 10	? × 100	? × 1,000
200	2 × 10	2 × 100	X
300			
5,000			
1,000			
700			
400			
100			
4,000			
800			
10,000			

Student Workbook • *Number Sense*　　Unit 3 Multiplication　**105**

PART B
DESIGNING A WEB PAGE

Activity: Look in the **Student Textbook** for the directions to design this Web page.

customer service
special features
more shopping
other stores

search

AND THERE'S MORE.....

ESTIMATION IN MULTIPLICATION
Problem Solving: Finding Important Information

LESSON 46

Name _____

PART A
WARM-UP ACTIVITY: BASIC AND EXTENDED FACTS

Activity: Solve the following basic and extended facts. Write the answer over the b.

1. 70 = 7 * ____
2. 600 = 6 * ____
3. 400 = ____ * 10
4. 700 = 70 * ____
5. 800 = 80 * ____
6. 70 = 7 * ____
7. 400 = 4 * ____
8. 400 = 40 * ____
9. 20 = ____ * 10
10. 80 = ____ * 10
11. 800 = 80 * ____
12. 800 = 8 * ____

PART B
ESTIMATION AND CALCULATORS

Activity: Use estimation and your calculator to solve the following problems.

1.   79
  × 43

   Estimation _____

   Calculator Answer ____

2.   48
  × 24

   Estimation _____

   Calculator Answer ____

3.   21
  × 89

   Estimation _____

   Calculator Answer ____

Student Workbook • Number Sense   Unit 3 Multiplication

ESTIMATING AND USING CALCULATORS
Designing a Logo for the Prism Company

LESSON 47

Name _____

PART A
WARM-UP ACTIVITY: BASIC AND EXTENDED FACTS

Activity: Solve the following basic and extended facts.

1. 6 × 9 = ____
2. 5 × 7 = ____
3. 9 × 6 = ____
4. 7 × 5 = ____

5. 6 × 90 = ____
6. 5 × 70 = ____
7. 9 × 60 = ____
8. 7 × 50 = ____

9. 60 × 9 = ____
10. 50 × 7 = ____
11. 90 × 6 = ____
12. 70 × 5 = ____

PART B
ESTIMATION AND CALCULATORS

Activity: Use estimation and your calculator to solve the following problems.

1.
   ```
     62
   × 62
   ```
   Estimation _____

   Calculator Answer _____

2.
   ```
     38
   × 42
   ```
   Estimation _____

   Calculator Answer _____

3.
   ```
     73
   × 68
   ```
   Estimation _____

   Calculator Answer _____

Student Workbook • *Number Sense* — Unit 3 Multiplication — 109

PART C
DRAWING THE PRISM COMPANY LOGO

Activity: Look at the design in the textbook. Use the space below to make a design similar to it. Be sure to do the following:

1. Put **The Prism Company** at the top.

2. Create a design in which the order of the triangles and the sizes match the textbook design. That means that the smallest triangle should be on the left, the largest one next, and so forth.

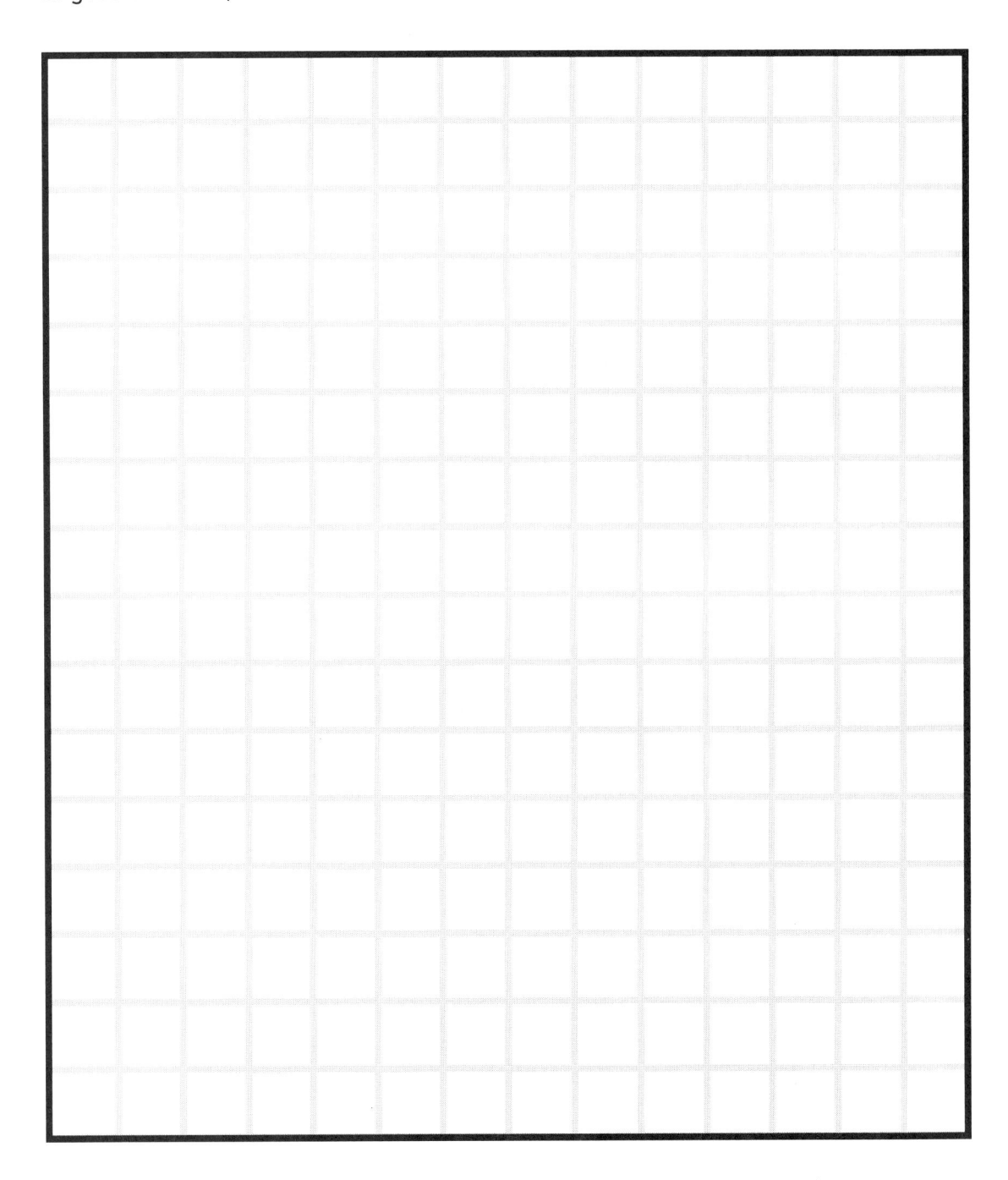

APPROXIMATING LARGE NUMBERS
Scale Drawings

LESSON 48

Name _____

PART A
WARM-UP ACTIVITY: BASIC AND EXTENDED FACTS

Activity: Solve the following basic and extended facts.

1. 10 × 10 = ____
2. 100 × 10 = ____
3. 90 × 10 = ____
4. 6 × 40 = ____
5. 100 × 10 = ____
6. 10 × 70 = ____
7. 90 × 10 = ____
8. 70 × 5 = ____
9. 10 × 10 = ____
10. 100 × 10 = ____
11. 9 × 60 = ____
12. 7 × 50 = ____

PART B
ESTIMATION AND CALCULATORS

Activity: Use estimation and your calculator to solve the following problems.

1. 203
   × 32

   Estimation _____
   Calculator Answer ____

2. 581
   × 43

   Estimation _____
   Calculator Answer ____

3. 333
   × 17

   Estimation _____
   Calculator Answer ____

4. 720
   × 30

   Estimation _____
   Calculator Answer ____

5. 342
   × 58

   Estimation _____
   Calculator Answer ____

Student Workbook • Number Sense — Unit 3 Multiplication — 111

PUTTING IT ALL TOGETHER
How Many Times Bigger?

LESSON 49

Name _____

PART A
WARM-UP ACTIVITY: BASIC AND EXTENDED FACTS

Activity: Solve the following basic and extended facts. Write the answer over the b.

1. ____ × 10 = 1,000
2. ____ × 10 = 100
3. 70 × ____ = 700
4. 7 × 50 = ____

5. 100 × 10 = ____
6. 10 × 90 = ____
7. ____ × 10 = 900
8. 60 × 5 = ____

9. 10 × ____ = 100
10. ____ × 10 = 1,000
11. 40 × 5 = ____
12. 7 × 50 = ____

PART B
MANY DIFFERENT WAYS

Activity 1: Find the exact answer to Problems 1 and 2 using the traditional method. Then find the exact answer to Problems 3 and 4 using the expanded method for multiplication.

Traditional Method:

1. 67
   × 7

2. 54
   × 6

Expanded Method:

3. 78
   × 5

4. 92
   × 8

Student Workbook • Number Sense   Unit 3 Multiplication   113

Activity 2: Use estimation and your calculator to solve the following problems.

1.   38
  × 32

    Estimation _____

    Calculator Answer _____

2.   604
  × 46

    Estimation _____

    Calculator Answer _____

3.   218
  × 31

    Estimation _____

    Calculator Answer _____

4.   63
  × 26

    Estimation _____

    Calculator Answer _____

5.   664
  × 12

    Estimation _____

    Calculator Answer _____

PART C
MORE ESTIMATIONS

1. About how long is a paper clip in real life? *(not this drawing)*

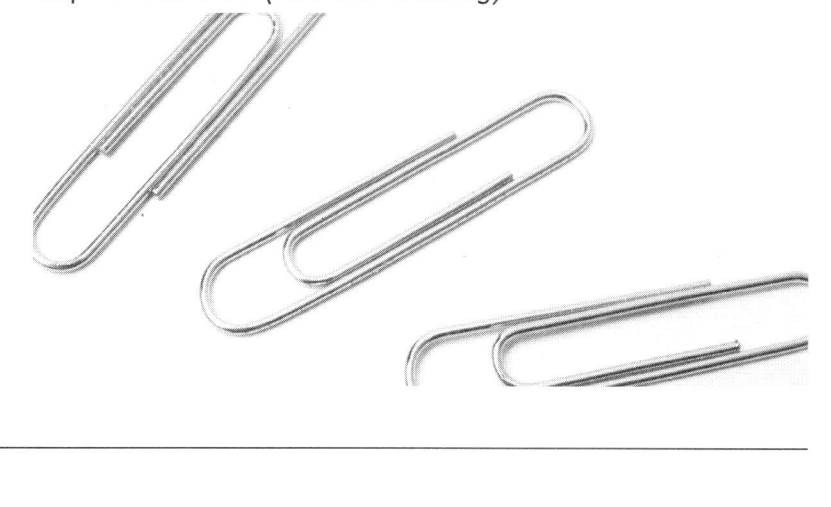

a. 5 mm

b. 5 cm

c. 50 cm

Tell why:

_____

_____

_____

2. About how long is a stapler in real life? *(not this drawing)*

a. 10 m

b. 20 cm

c. 500 cm

Tell why:

_____

_____

_____

Student Workbook • *Number Sense* Unit 3 Multiplication • Lesson 49 **115**

3. About how long is a pencil in real life? *(not this drawing)*

a. 15 mm

b. 15 cm

c. 15 meters

Tell why:

_____

_____

_____

UNIT 3 REVIEW

Name _____

LESSON 50

PART A
COMMUTATION IN MULTIPLICATION

Activity: Solve the following turn-around multiplication facts.

1. $7 \times 8 = 8 \times$ ____ $= 56$
2. $5 \times 6 = 6 \times 5 =$ ____
3. $9 \times$ ____ $= 6 \times 9 = 54$
4. $4 \times 8 = 8 \times$ ____ $=$ ____

PART B
BASIC AND EXTENDED MULTIPLICATION FACTS

Activity: Solve the following basic and extended multiplication facts.

1. $7 \times 9 =$ ____
2. $6 \times 8 =$ ____
3. $7 \times 90 =$ ____
4. $6 \times 80 =$ ____
5. $7 \times 900 =$ ____
6. $6 \times 800 =$ ____

PART C
PULLING OUT THE 10

Activity: Fill in the table by writing the following numbers as "something $\times$ 10."

Starting Number	? $\times$ 10
50	5 $\times$ 10
60	
70	
80	
90	

Student Workbook • *Number Sense* Unit 3 Multiplication 117

PART D
EXPANDED FORM

Activity: Solve the following multiplication problems using expanded form.

1. 37
    × 4

    × ___|___

2. 98
    × 6

    × ___|___

PART E
FINDING MULTIPLES OF 10

Activity: Fill in the table by writing the number as something times 10, something times 100, and something times 1,000.

Starting Number	? × 10	? × 100	? × 1,000
7,000			
8,000			
9,000			

PART F
TRADITIONAL MULTIPLICATION

Activity: Solve using traditional multiplication.

1. 48
   × 5

2. 54
   × 8

3. 96
   × 7

4. 87
   × 9

PART G
ESTIMATION IN MULTIPLICATION

Activity: Estimate the answers to the following problems by rounding the numbers and solving the extended fact.

1. 27
   × 4

   Estimated Extended Fact _____

2. 61
   × 37

   Estimated Extended Fact _____

3. 419
   × 78

   Estimated Extended Fact _____

PART H
COMMON ERRORS IN MULTIPLICATION

Activity: Each of the following problems contains an error. Use estimation and a calculator to find the correct answer. Then write a sentence describing the error.

1. 79
    $\times$3
    ─────
    217

2. 48
    $\times$ 14
    ─────
    192
    + 48
    ─────
    240

3. 96
    $\times$7
    ─────
    654

4. 75
    $\times$9
    ─────
    684

1. Estimation _____

   Calculator Answer ____

   _____

   _____

2. Estimation _____

   Calculator Answer ____

   _____

   _____

3. Estimation _____

   Calculator Answer ____

   _____

   _____

4. Estimation _____

   Calculator Answer ____

   _____

   _____

PART I
METRIC MEASUREMENT

1. About how long is a baseball bat in real life? *(not this drawing)*

   a. 10 millimeters

   b. 1 centimeter

   c. 1 meter

   Tell why:

   _____

   _____

2. About how long are the skis in real life? *(not this drawing)*

   a. 2 meters

   b. 12 meters

   c. 20 meters

   Tell why:

   _____

   _____

   _____

3. About how long is the ship in real life? *(not this drawing)*

   a. 10 meters

   b. 100 meters

   c. 100 centimeters

   Tell why:

   _____

   _____

MULTIPLICATION AND DIVISION FACT FAMILIES

Measuring in Square Units

Name _____

LESSON 51

PART A
WARM-UP ACTIVITY: BASIC DIVISION FACTS

Activity: Solve the following basic facts.

1. 81 ÷ 9 = _____
2. 21 ÷ 7 = _____
3. 42 ÷ 6 = _____
4. 24 ÷ 4 = _____
5. 35 ÷ 5 = _____
6. 56 ÷ 8 = _____
7. 45 ÷ 9 = _____
8. 18 ÷ 3 = _____

PART B
MULTIPLICATION/DIVISION FACT FAMILIES

Activity 1: Write fact families for the following groups of numbers.

EXAMPLE

9, 8, and 72

9 × 8 = 72

8 × 9 = 72

72 ÷ 9 = 8

72 ÷ 8 = 9

1. 6, 7, and 42

3. 8, 7, and 56

5. 3, 9, and 27

2. 5, 7, and 35

4. 8, 9, and 72

6. 3, 6, and 18

Student Workbook • *Number Sense* Unit 4 Division **123**

Activity 2: Solve the following word problems. Show the multiplication or division fact that you used to find the answer for each problem.

1. Spike is a fan of the top 8 teams in the Western Conference of the NBA. He follows teams like the LA Lakers and the Portland Trail Blazers. He wants to sort his basketball cards into equal piles with 5 cards in a pile. He has 40 cards. How many piles will he have?

2. Julia, Marie, and Inez also collect basketball cards of players on their favorite teams. They find that each of them has 9 Lakers cards. When they put all of the Lakers cards out on the table, how many do they have?

3. Seats near the court at any basketball playoff game are always hard to get. Suppose that there are only 6 rows of seats near the basketball court and there are only 3 open seats in each row. How many open seats are there in the 6 rows?

4. The home team wants to pass out towels for the fans to wave when their team is winning. They want to give the towels to people in the first 4 rows. There 45 towels. How many towels can they give to people in each row if they give the same number of towels to each row?

PART C
SQUARE UNITS

Activity: How many square units are there in this shape?

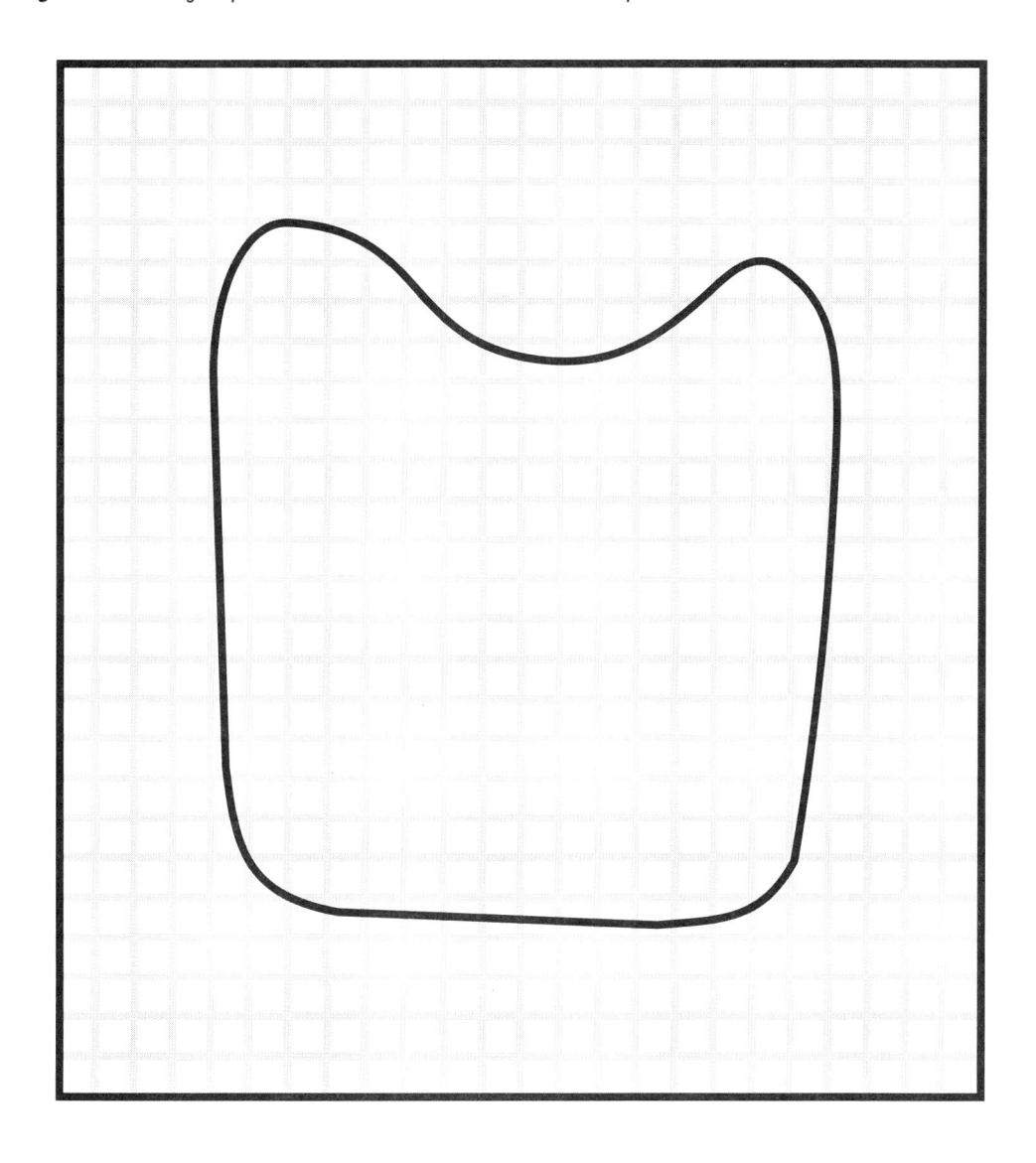

Explain how you got your answer: _____

_____

_____

Student Workbook • *Number Sense* Unit 4 Division • Lesson 51 **125**

BASIC AND EXTENDED DIVISION FACTS
Comparing Sizes of Shapes

Lesson 52

Name _____

PART A
WARM-UP ACTIVITY: FACT FAMILIES

Activity: Write fact families for the following sets of numbers.

1. 6, 4, and 24

2. 8, 6, and 48

3. 5, 9, and 45

PART B
EXTENDED FACTS IN DIVISION

Activity: Fill in the missing basic and extended facts in the table.

Basic Fact	10s Fact	100s Fact
54 ÷ 6 = 9	540 ÷ 6 = 90	5,400 ÷ 6 = 900
81 ÷ 9 = 9		8,100 ÷ 9 = 900
36 ÷ 4 = 9	360 ÷ 4 = 90	
		2,400 ÷ 4 = 600
72 ÷ 8 = 9		7,200 ÷ 8 = 900
	120 ÷ 4 = 30	

Student Workbook • *Number Sense*   Unit 4 Division   **127**

PART C
MORE DIVISION FACTS AND EXTENDED FACTS

Activity: Solve the following basic and extended division facts.

1. $9\overline{)27}$

2. $7\overline{)42}$

3. $3\overline{)240}$

4. $9\overline{)270}$

5. $8\overline{)72}$

6. $6\overline{)54}$

7. $5\overline{)400}$

8. $8\overline{)720}$

PART D
COMPARING SIZE

Activity: Which of these shapes is the largest?

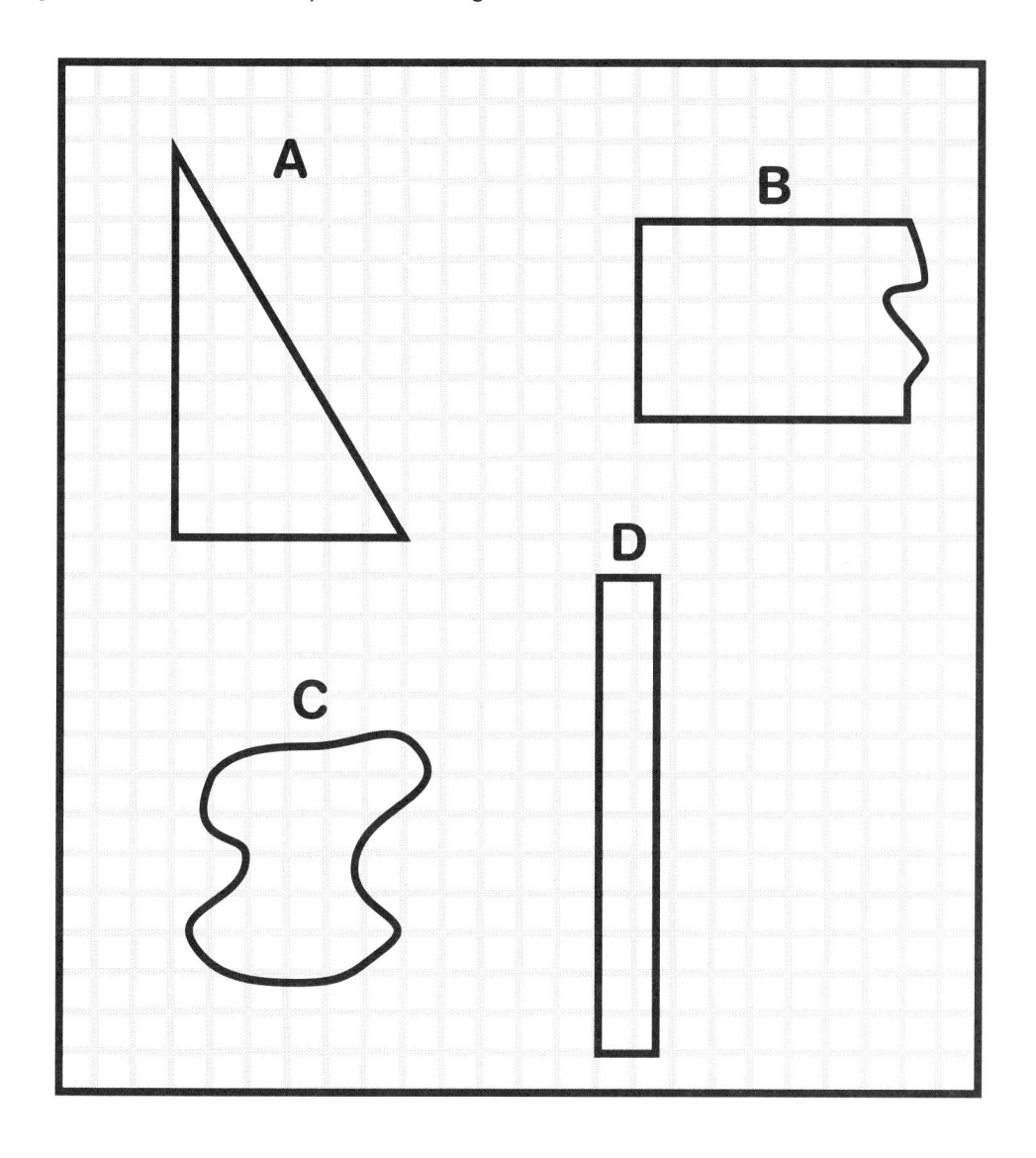

Tell how you got your answer.

Student Workbook • *Number Sense*

BASIC DIVISION FACTS ON A NUMBER LINE
Profits From May Concerts

Lesson 53

Name _____

PART A
WARM-UP ACTIVITY: BASIC AND EXTENDED FACTS

Activity: Solve the following basic and extended facts.

1. 9)18

2. 5)450

3. 7)490

4. 8)64

5. 6)240

6. 3)300

PART B
DIVISION ON THE NUMBER LINE

Activity: Solve the following basic facts and show how you got your answers on the number lines provided. Be sure to label the number lines with the best scale for each problem.

EXAMPLE

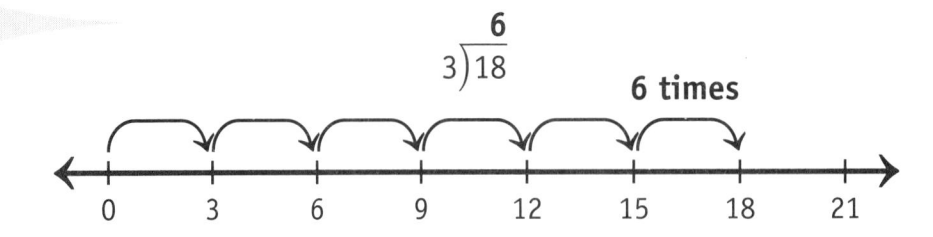

$$3\overline{)18} = 6$$

6 times

```
  0   3   6   9   12  15  18  21
```

1. $7\overline{)21}$

2. $6\overline{)24}$

3. $7\overline{)28}$

4. $3\overline{)12}$

5. $9\overline{)18}$

132   Unit 4 Division • Lesson 53   Student Workbook • *Number Sense*

EXTENDED DIVISION FACTS ON A NUMBER LINE
Same Shape, Same Size?

LESSON 54

Name _____

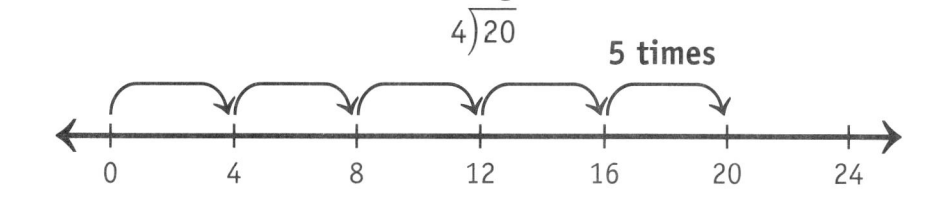

PART A
WARM-UP ACTIVITY: NUMBER LINES

Activity: Solve the following division facts and show how you got your answers on the number lines provided. Be sure to label the number lines with the best scale for each problem.

EXAMPLE

$4\overline{)20} = 5$, 5 times

0 4 8 12 16 20 24

1. $7\overline{)14}$

2. $8\overline{)24}$

3. $6\overline{)30}$

4. $9\overline{)36}$

Student Workbook • *Number Sense* Unit 4 Division **133**

PART B
EXTENDED FACTS IN DIVISION

Activity: Fill in the missing basic and extended facts in the table.

Basic Fact	10s Fact	100s Fact
72 ÷ 8 = 9	720 ÷ 8 = 90	7,200 ÷ 8 = 900
49 ÷ 7 = 7		4,900 ÷ 7 = 700
		2,500 ÷ 5 = 500
	240 ÷ 6 = 40	
42 ÷ 7 = 6		
		2,700 ÷ 9 = 300
		1,600 ÷ 4 = 400

PART C
SAME SHAPE, SAME SIZE?

Activity: Find the area of Object A and Object B. Remember to answer using *square units* by counting the square units for each one.

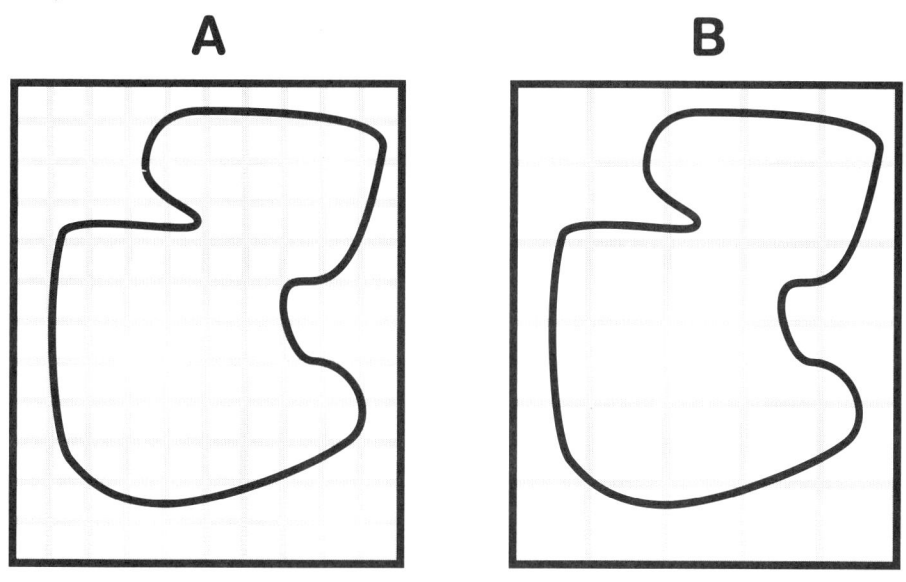

A

B

Area of Object A: _____

Area of Object B: _____

Tell why the measurements for the two objects are different.

_____

_____

_____

_____

Student Workbook • *Number Sense*   Unit 4 Division • Lesson 54   **135**

REMAINDERS
Estimating Square Units on a Map

Lesson 55

Name _____

PART A
WARM-UP ACTIVITY: BASIC AND EXTENDED FACTS

Activity: Solve the following basic and extended facts.

1. $6\overline{)18}$

2. $9\overline{)450}$

3. $4\overline{)24}$

4. $6\overline{)360}$

PART B
REMAINDERS

Activity 1: Solve the following problems using your calculator. Then answer the questions that follow the problems.

1. $4\overline{)34}$

   a. How many equal units of 4 are in 34? _____

   b. What's the decimal number remainder? _____

2. $8\overline{)52}$

   a. How many equal units of 8 are in 52? _____

   b. What's the decimal number remainder? _____

Student Workbook • *Number Sense* — Unit 4 Division — 137

Activity 2: Use what you know about division facts and their remainders to solve the following problems. Then, on the number line, draw a line to show the number you are dividing up. Then circle the remainder as shown in the example.

EXAMPLE

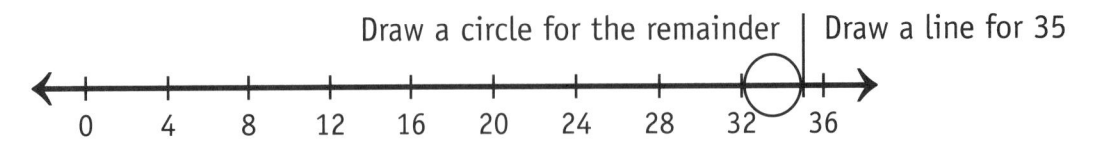

$$4\overline{)35} = 8 \text{ R}3$$

Draw a circle for the remainder | Draw a line for 35

←―+―+―+―+―+―+―+―+―+―→
 0 4 8 12 16 20 24 28 32 36

1. $3\overline{)23}$

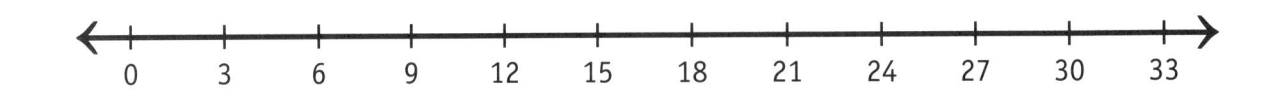

←―+―+―+―+―+―+―+―+―+―+―+―→
 0 3 6 9 12 15 18 21 24 27 30 33

2. $6\overline{)41}$

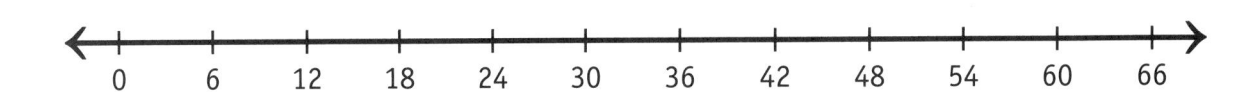

←―+―+―+―+―+―+―+―+―+―+―+―→
 0 6 12 18 24 30 36 42 48 54 60 66

3. $7\overline{)45}$

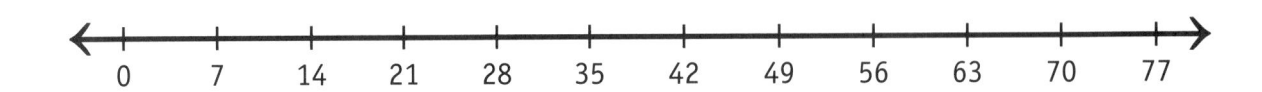

←―+―+―+―+―+―+―+―+―+―+―+―→
 0 7 14 21 28 35 42 49 56 63 70 77

4. $5\overline{)24}$

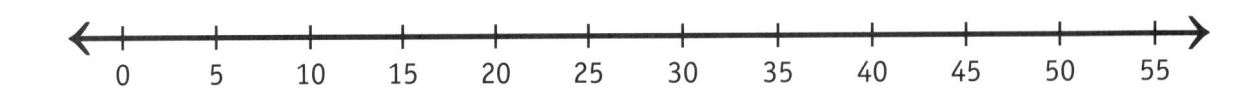

←―+―+―+―+―+―+―+―+―+―+―+―→
 0 5 10 15 20 25 30 35 40 45 50 55

PART C
THE POPULATION OF SAN FRANCISCO

Activity: Here is a map of the city of San Francisco, California. San Francisco is one of the most crowded cities in the United States. As of 1997, there were 32,471 housing units in the city. The land area of San Francisco is 46.7 square miles.

Each *full* square below contains about 26,400 people. Count the squares to figure out the approximate population of San Francisco. Note that not all squares are full squares, so you are going to have to combine partial squares to make full squares.

What is the approximate population of San Francisco? _____

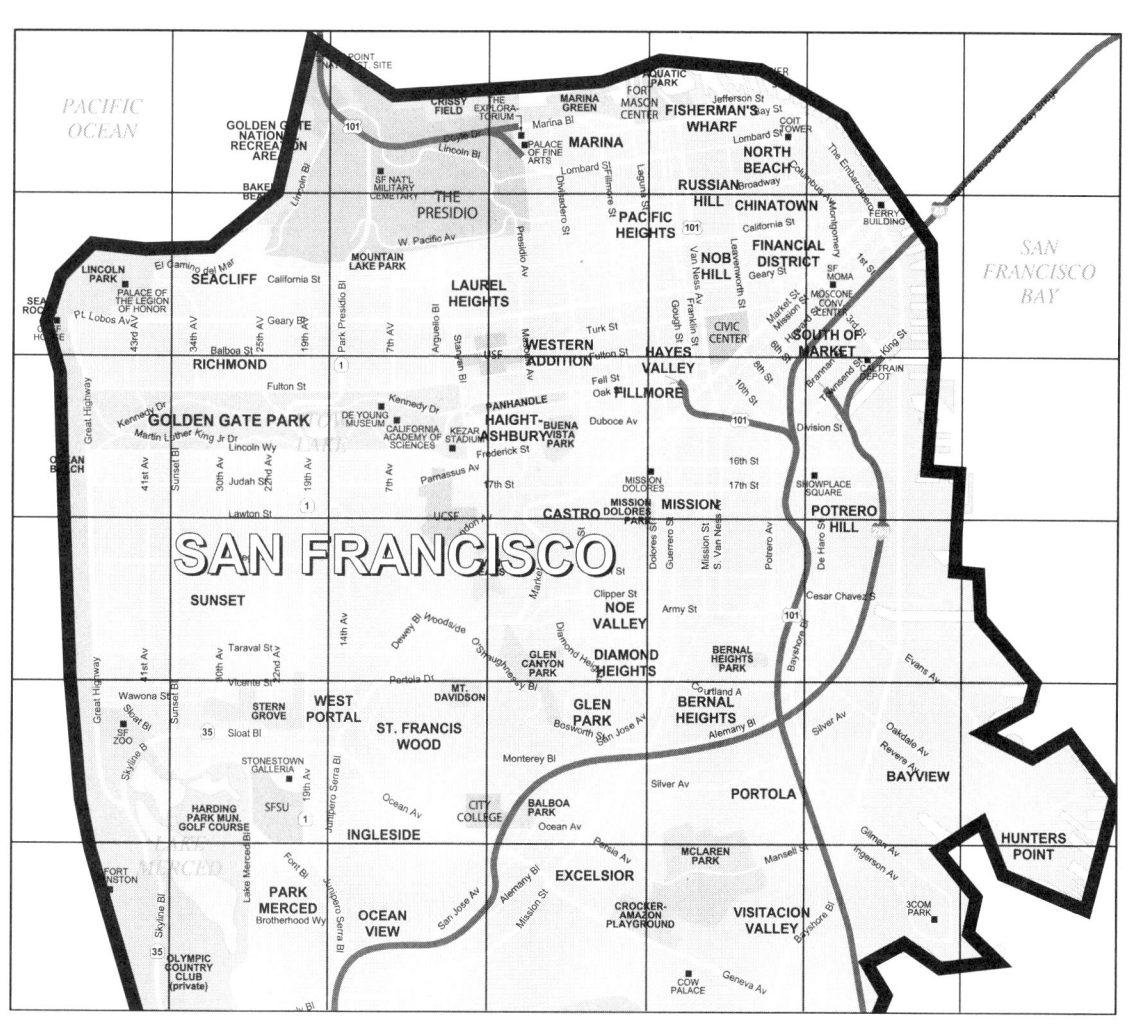

Tell how you got your answer.

_____
_____
_____

Student Workbook • *Number Sense* Unit 4 Division • Lesson 55 **139**

USING CALCULATORS FOR DIVISION
Rounding Strategies in Division

LESSON 56

Name _____

PART A
WARM-UP ACTIVITY: DIVISION WITH A CALCULATOR

Activity: Use your calculator to solve the following problems. Be sure to write the entire answer that is on your calculator.

1. 6)19

2. 9)28

3. 4)25

4. 6)37

PART B
ROUNDING TO THE NEAREST WHOLE NUMBER

Activity: Use your calculator to solve the following problems. Show the calculator answers. Then round each answer to the nearest whole number.

1. 8)34

    Calculator Answer: _____

    Rounded Answer: _____

2. 7)58

    Calculator Answer: _____

    Rounded Answer: _____

3. 5)41

    Calculator Answer: _____

    Rounded Answer: _____

4. 3)16

    Calculator Answer: _____

    Rounded Answer: _____

Student Workbook • *Number Sense* Unit 4 Division

"NEAR FACT" DIVISION
Finding Area By Counting Squares

LESSON 57

Name _____

PART A
WARM-UP ACTIVITY: DIVISION WITH A CALCULATOR

Activity: Use your calculator to solve the following division problems. Then round to the nearest whole number.

1. 6)13

   Calculator Answer: ____

   Rounded Answer: ____

2. 9)85

   Calculator Answer: ____

   Rounded Answer: ____

3. 8)50

   Calculator Answer: ____

   Rounded Answer: ____

4. 3)19

   Calculator Answer: ____

   Rounded Answer: ____

PART B
NEAR FACTS

Activity: Estimate the answers to the following problems by finding the closest basic fact (the near fact). Then solve the original problem with a calculator to see how good your estimation is.

EXAMPLE

8)57

Near Fact: 8)56 with 7 on top

1. 6)13

   Near Fact: ____)____

   Calculator Answer: ____

2. 9)85

   Near Fact: ____)____

   Calculator Answer: ____

3. 7)44

   Near Fact: ____)____

   Calculator Answer: ____

4. 8)34

   Near Fact: ____)____

   Calculator Answer: ____

Student Workbook • Number Sense

Unit 4 Division

5. 4)31̄

   Near Fact: ___)___

   Calculator Answer: ___

6. 5)34̄

   Near Fact: ___)___

   Calculator Answer: ___

7. 2)19̄

   Near Fact: ___)___

   Calculator Answer: ___

8. 9)75̄

   Near Fact: ___)___

   Calculator Answer: ___

PART C
SNOWBOARD DESIGN

Activity: On the next page you will see a star shaped design. The design is going to be used on the top of a snowboard. Artists make drawings like this before they use them on a finished product to see what they look like and to find out what other people think about them.

Answer the following questions about the star shaped design. Here's a hint. You don't have to count every square to figure out the size of each object. Give your answers in square units.

For each question, round your answer to the nearest 10.

1. About how big is the small star in the center? ___

2. About how big is the larger star? ___

3. About how much bigger is the larger star than the smaller star? (Be sure to include the size of the smaller star *inside* of the larger star when you figure this out). ___

4. About how big is the whole space inside the circle? ___

5. In the space below, describe the strategy that you used to figure out the size of each of these objects.

   _____
   _____
   _____
   _____
   _____
   _____

Snowboard Design

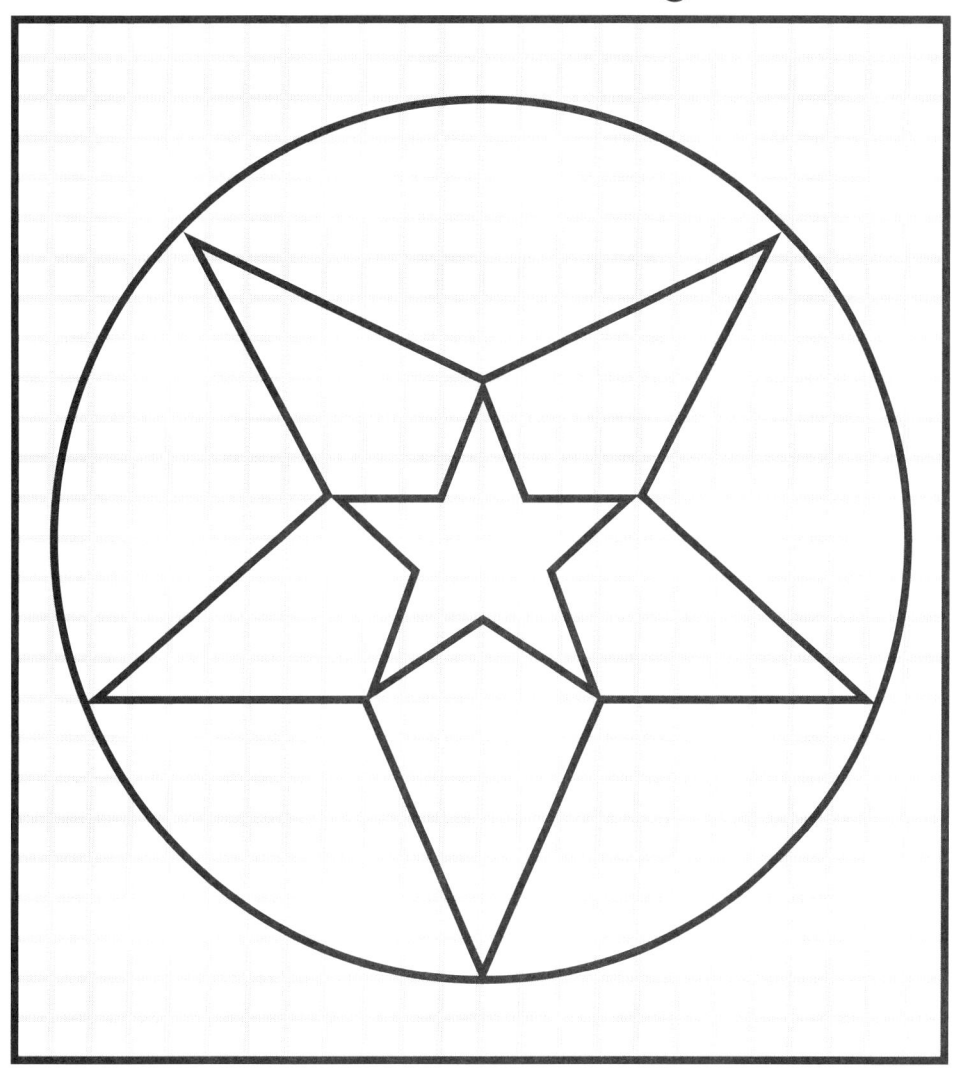

REVIEW OF LESSONS 51 TO 57

Name _____

Lesson 58

PART A
BASIC DIVISION FACTS

Activity: Solve the following basic division facts.

1. 56 ÷ 8 = _____
2. 49 ÷ _____ = 7
3. _____ ÷ 9 = 8

PART B
EXTENDED DIVISION FACTS

Activity: Solve the following extended division facts.

1. 560 ÷ 8 = _____
2. 490 ÷ 7 = _____
3. 810 ÷ 9 = _____

PART C
FACT FAMILIES

Activity: Write fact families for the following groups of numbers. Replace the X with the correct number.

1. 7, 4, and 28
2. 8, 7, and X
3. X, 3, and 12

_____   _____   _____

_____   _____   _____

_____   _____   _____

_____   _____   _____

PART D
NUMBER LINES

Activity: Solve the following basic facts. Show how you got your answers on the number lines provided.

1. 7)28

⟵—————————————————⟶

2. 8)24

⟵—————————————————⟶

PART E
DIVISION ON THE CALCULATOR

Activity: Use your calculator to solve the following problems with remainders. Round the answers to the nearest whole number.

1. 9)28

   Calculator Answer: _____

   Rounded Answer: _____

2. 8)59

   Calculator Answer: _____

   Rounded Answer: _____

3. 7)37

   Calculator Answer: _____

   Rounded Answer: _____

4. 4)38

   Calculator Answer: _____

   Rounded Answer: _____

PART F
NEAR FACTS

Activity: Estimate the answers to the following problems by finding the closest basic fact (the near fact). Then solve the original problems with a calculator to see how good your estimate is.

$$8\overline{)57}$$

Near Fact: $8\overline{)56}^{\,7}$

EXAMPLE

1. $6\overline{)47}$

    Near Fact: _____

    Calculator Answer: _____

2. $9\overline{)78}$

    Near Fact: _____

    Calculator Answer: _____

Student Workbook • *Number Sense*   Unit 4 Division • Lesson 58

DIVISION: THE TRADITIONAL ALGORITHM
Careers in Architecture

LESSON 59

Name _____

PART A
WARM-UP ACTIVITY: BASIC AND NEAR FACTS

Activity 1: Solve the following basic facts.

1. 27 ÷ 9 = ____
2. 24 ÷ 6 = ____
3. 42 ÷ 7 = ____
4. 18 ÷ 3 = ____
5. 35 ÷ 7 = ____
6. 56 ÷ 8 = ____
7. 63 ÷ 9 = ____
8. 30 ÷ 3 = ____

Activity 2: Rewrite the following problems as near facts.

1. 7)47
   Near Fact: ____)____

2. 9)29
   Near Fact: ____)____

3. 4)37
   Near Fact: ____)____

4. 8)51
   Near Fact: ____)____

PART B
TRADITIONAL LONG DIVISION

Activity: Solve the following problems using long division.

1. 8)564
2. 5)158
3. 7)701
4. 7)144
5. 3)210
6. 9)814

PART C
FLOOR PLANS

Activity: Architects make blueprint drawings before they build a house. These drawings are very important for many reasons. One reason is that builders use these plans to figure out the types and amount of materials they need to build the house.

The blueprint drawing on the next page shows the design of the bottom floor of a house. You can see two bedrooms, a kitchen, a bathroom, and a living room. You can also see lines for walls, windows, and doors.

Answer the following questions based on the blueprint drawing. Hint: When you figure out how many square units a room is, don't worry about the doors on the inside of the room. Similarly, when you figure out the size of the inside of the house, don't worry about the doors, inside walls, or windows.

1. How big is the inside of the house including all of the rooms? _____

2. Draw lines so that each room is a rectangle. This will be especially important for the kitchen and the living room.

    a. Which room is the biggest? _____

    b. Which room is the smallest? _____

    c. How much bigger is the kitchen than the bathroom? _____

    d. How much bigger is bedroom A than bedroom B? _____

bedroom A

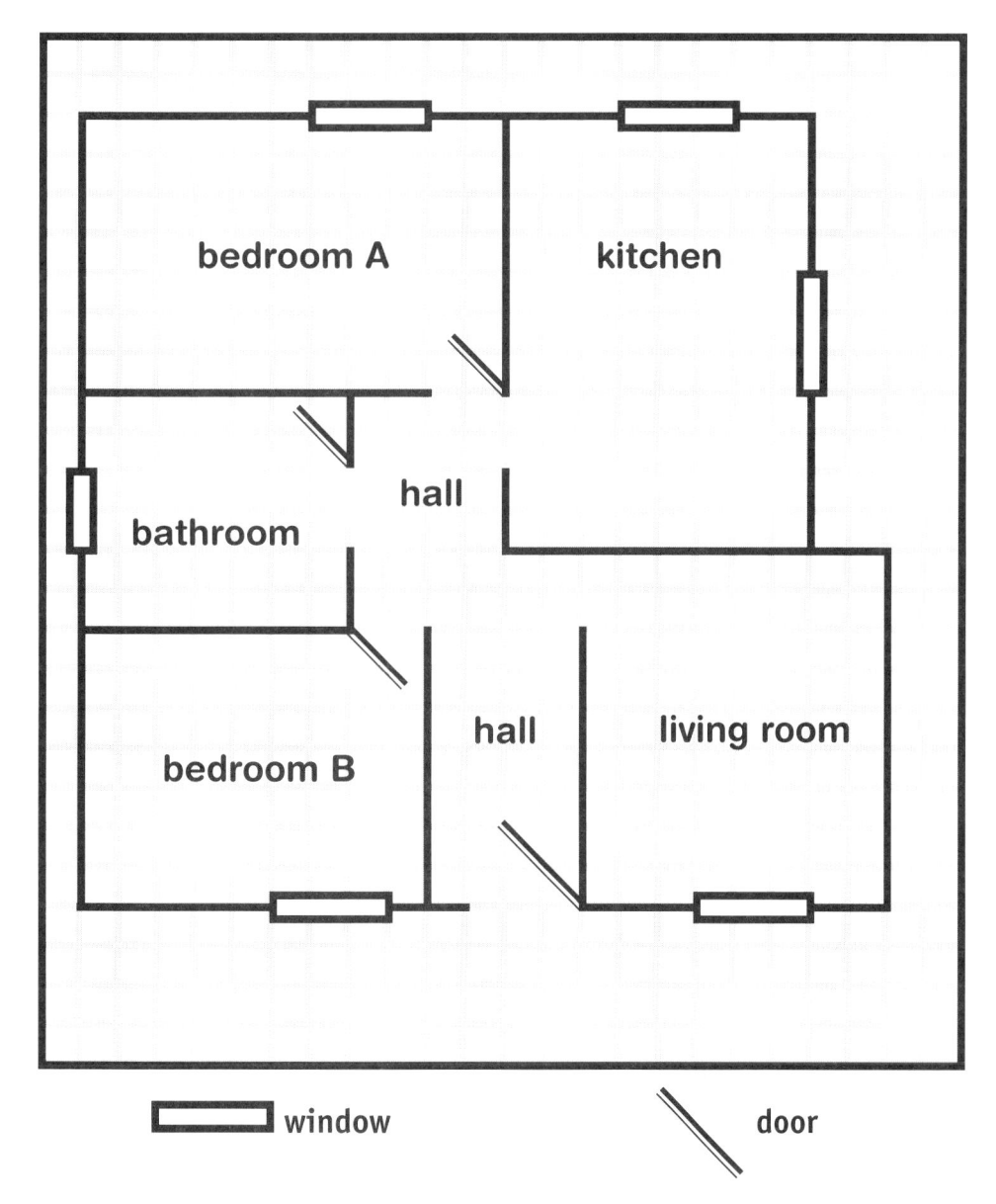

kitchen

hall

bathroom

bedroom B

hall

living room

window   door

Student Workbook • *Number Sense*   Unit 4 Division • Lesson 59   153

COMPARING DIVISION ALGORITHMS
Estimating Areas in Floor Plans

LESSON 60

Name _____

PART A
WARM-UP ACTIVITY: LONG DIVISION

Activity 1: Fill in the missing basic and extended facts in the table.

Basic Fact	10s Fact	100s Fact
54 ÷ 6 = 9	540 ÷ 6 = 90	5,400 ÷ 6 = 900
64 ÷ 8 = 8		6,400 ÷ 8 = 800
36 ÷ 9 = 4	360 ÷ 9 = 40	
		1,800 ÷ 9 = 200

Activity 2: Solve the following problems using long division.

1. 7)̄413

2. 8)̄592

Student Workbook • *Number Sense*   Unit 4 Division   155

PART B
NEAR EXTENDED FACTS

Activity: Rewrite the following problems using near extended facts. Do not solve the original problems.

EXAMPLE $8\overline{)562}$

Near Extended Fact: $8\overline{)560}^{\,70}$

1. $6\overline{)122}$

   Near Extended Fact: $\overline{)}$

2. $9\overline{)178}$

   Near Extended Fact: $\overline{)}$

3. $4\overline{)203}$

   Near Extended Fact: $\overline{)}$

4. $5\overline{)356}$

   Near Extended Fact: $\overline{)}$

5. $7\overline{)488}$

   Near Extended Fact: $\overline{)}$

PART C
BEDROOM FURNITURE

Activity: One thing that an architect often adds to a room design to show what the room will look like is furniture. The drawing on the next page has a bed, a closet, a rug, and a desk.

Answer the following questions about the room floor plan.

1. What is the size of each object in square units?

   Bed ____

   Closet ____

   Rug ____

   Desk ____

2. What is the total size of the bedroom in this drawing? (Be sure to include all of the objects.)

3. Suppose you want to put *another* bed in this room that is 9 square units wide and 12 square units long. That means there would be two beds. Is there enough room to put a second bed in the room so that it does not cover any part of the rug without moving the furniture already in the room?

bed

closet

rug

desk

window   door

Student Workbook • *Number Sense*   Unit 4 Division • Lesson 60   **157**

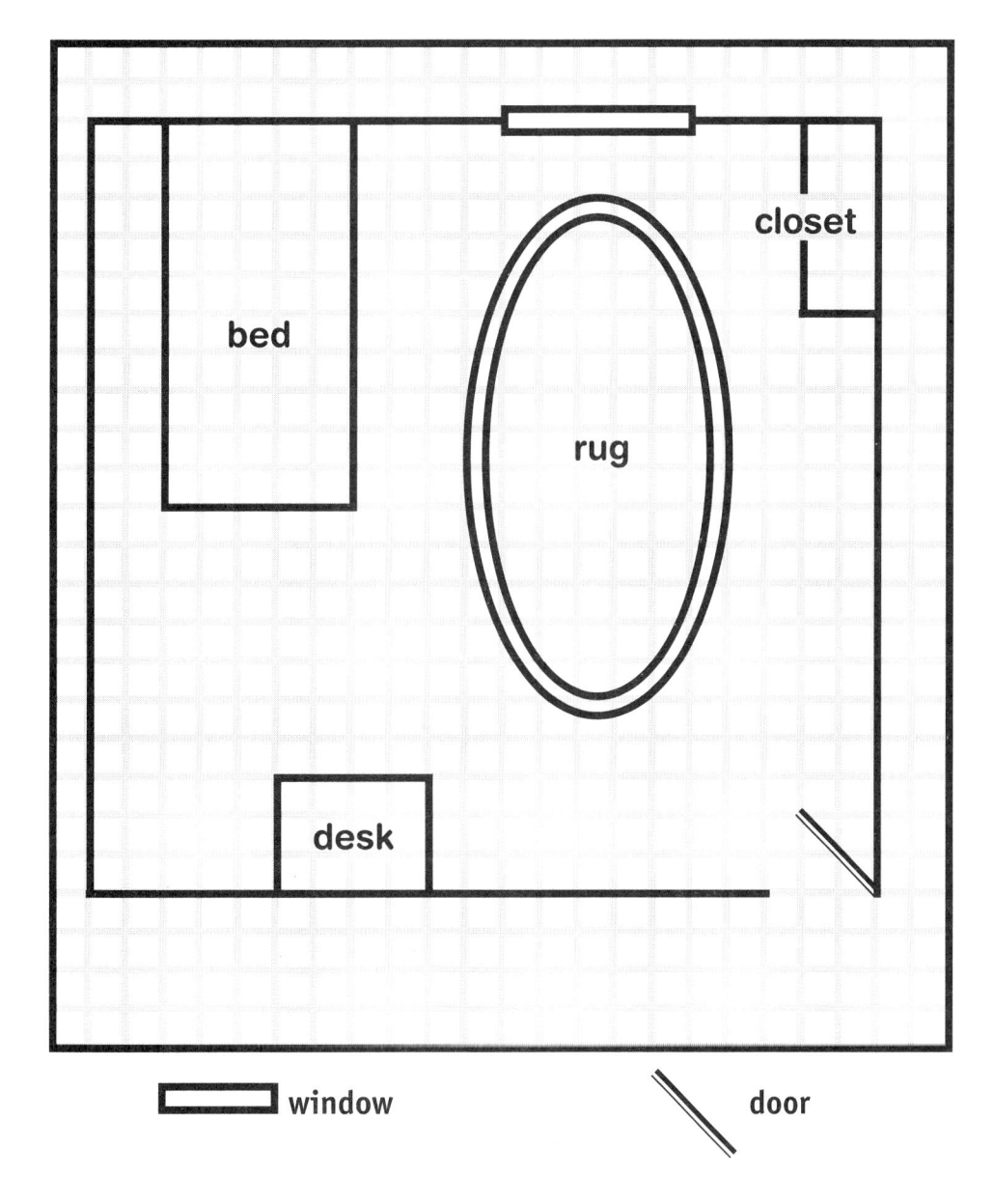

LOOKING AT BIGGER DIVISION PROBLEMS
More Floor Plans

LESSON 61

Name _____

PART A
WARM-UP ACTIVITY: NEAR EXTENDED FACTS AND FACT FAMILIES

Activity 1: Rewrite the following problems as near extended facts. Do not solve the original problems.

1. $7\overline{)489}$

    Near Fact: _____

2. $6\overline{)419}$

    Near Fact: _____

Activity 2: Write fact families for the following groups of numbers.

1. 6, 7, and 42

2. 9, 3, and 27

3. 6, 9, and 54

_____   _____   _____

_____   _____   _____

_____   _____   _____

_____   _____   _____

PART B
ESTIMATIONS WHEN DIVIDING BY TWO-DIGIT NUMBERS

Activity: Rewrite both numbers to make a near extended fact. Do not solve the original problems.

$91\overline{)371}$

Near Extended Fact: $90\overline{)360}$ (with 4 on top)

EXAMPLE

1. $82\overline{)564}$

    Near Extended Fact: _____

2. $61\overline{)397}$

    Near Extended Fact: _____

3. $49\overline{)279}$

    Near Extended Fact: _____

4. $38\overline{)406}$

    Near Extended Fact: _____

PART C
PULLING OUT THE 10s

Activity: Rewrite the following near extended facts by pulling out the 10s. Do not solve the original problem or the near extended fact.

EXAMPLE

$60 \overline{)240}$

Tens: $6 \times 10 \overline{)24 \times 10}$

1. $90 \overline{)360}$

   Tens: $\overline{)}$

2. $70 \overline{)490}$

   Tens: $\overline{)}$

3. $60 \overline{)240}$

   Tens: $\overline{)}$

4. $80 \overline{)560}$

   Tens: $\overline{)}$

5. $30 \overline{)90}$

   Tens: $\overline{)}$

PART D
DESIGN YOUR OWN LIVING ROOM

Activity: Here's a chance to design your own living room. Use the space on the next page. Put at least four objects in the room. They could be chairs, couches, a TV, shelves, a coffee table, or anything else that might go in the living room.

Before you draw the objects, think about how big they should be, and write this information in the table on this page. Then draw the objects on the floor plan. Be sure to label.

Name of the Object	Number of Square Units

Design Your Own Living Room

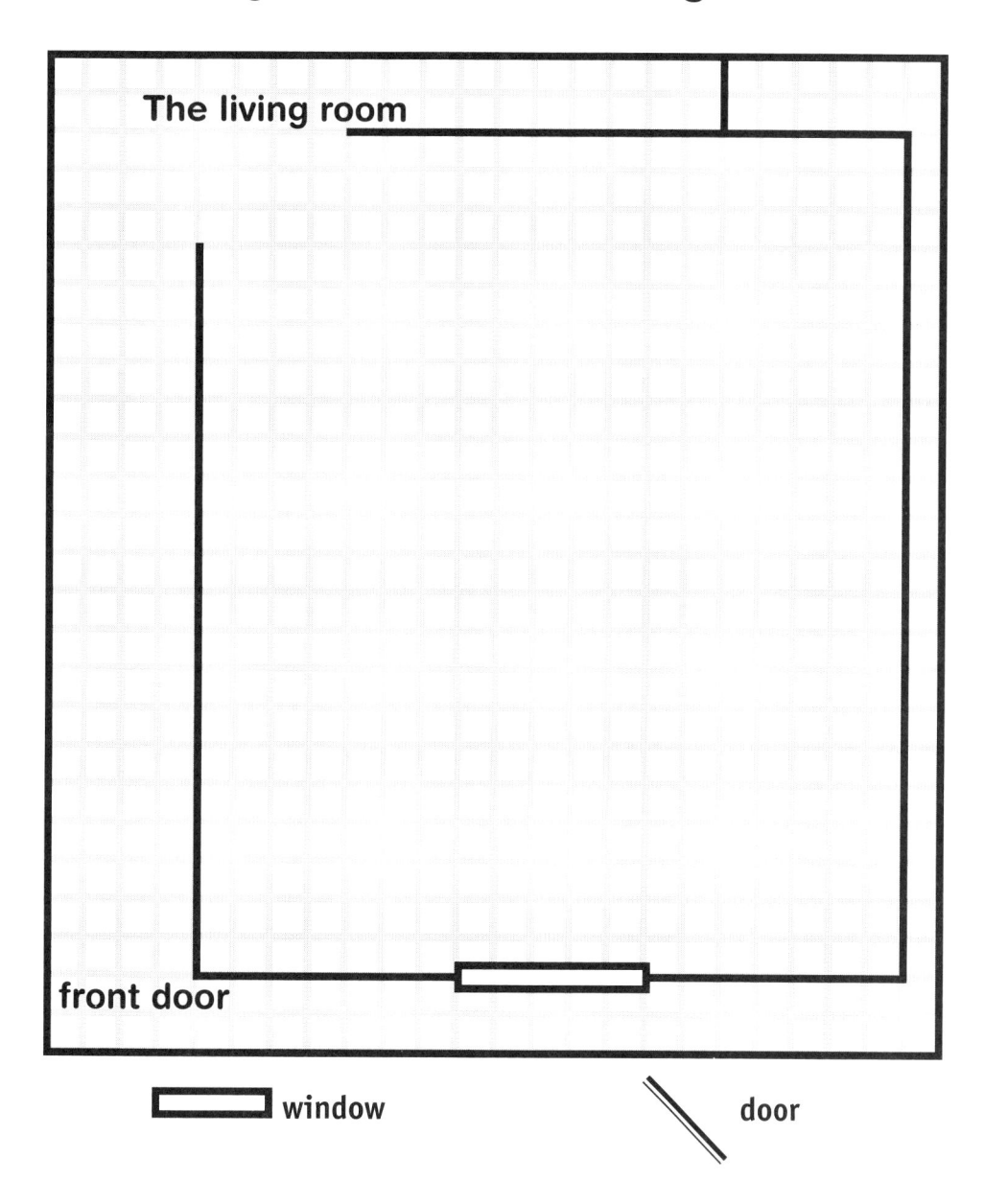

The living room

front door

☐▭☐ window   ╱╱ door

PULLING OUT THE 10s
Problem Solving: The Arena Building

Name _____

PART A
WARM-UP ACTIVITY: PULLING OUT THE 10s

Activity 1: Find the missing values in the following division facts. B stands for a missing value.

1. 81 ÷ ____ = 9
2. 21 ÷ ____ = 3
3. 48 ÷ ____ = 6
4. 24 ÷ ____ = 6
5. 30 ÷ ____ = 5
6. 72 ÷ ____ = 8

Activity 2: Rewrite the following near extended facts by pulling out the 10s. Do not solve.

1. 90)810

    Pull Out the 10s: ____)____

2. 70)210

    Pull Out the 10s: ____)____

3. 80)720

    Pull Out the 10s: ____)____

4. 60)480

    Pull Out the 10s: ____)____

PART B
10s AND NEAR EXTENDED FACTS

Activity: Fill in the missing items in the table. You are asked to rewrite the near extended facts by pulling out the 10s. Then you are to rewrite the problem as a basic fact and solve it. Finally, you are to solve the near extended fact.

Near Extended Fact	Pull Out the 10s	Basic Fact	Near Extended Fact and Solution
$60\overline{)240}$	$6 \times 10\overline{)24 \times 10}$	$6\overline{)24}$ = 4	$60\overline{)240}$ = 4
$70\overline{)280}$			
$30\overline{)210}$			
$50\overline{)250}$			
$20\overline{)180}$			
$80\overline{)320}$			

Unit 4 Division • Lesson 62 — Student Workbook • *Number Sense*

IDENTIFYING MISTAKES IN DIVISION ON THE CALCULATOR

Problem Solving: Making Division Word Problems

Name _____

LESSON 63

PART A
WARM-UP ACTIVITY: BASIC FACTS AND PULLING OUT THE 10s

Activity 1: Find the missing value in the following division facts. K stands for a missing value.

1. 64 ÷ ____ = 8
2. 12 ÷ ____ = 2
3. 28 ÷ ____ = 4
4. 15 ÷ ____ = 5
5. 63 ÷ ____ = 7
6. 60 ÷ ____ = 6

Activity 2: Rewrite the following near extended facts by pulling out the 10s Then solve the basic fact.

$$30 \overline{)180} \qquad 3 \times 10 \overline{)18 \times 10} \qquad 3 \overline{)18}^{\,6} \qquad \text{EXAMPLE}$$

1. $20 \overline{)120}$

2. $80 \overline{)640}$

3. $40 \overline{)280}$

4. $70 \overline{)630}$

Student Workbook • *Number Sense* — Unit 4 Division — 165

PART B
COMMON CALCULATOR ERRORS

Activity: A student used a calculator to solve the following problem.

$$7\overline{)31}$$

She got this answer on her calculator:

.2258064516129

1. Estimate the answer by finding a near fact.
2. Find the exact answer on a calculator.
3. Describe the error the student made.

_____

_____

_____

PART C
DOING LONG DIVISION BY HAND: FIND THE ERROR

Activity: This time a student did three long division problems by hand. He made *at least* one mistake. Estimate the answer to each problem first. Then find which problem or problems are incorrect.

A.
```
      20 R3
   7)143
    -14
     03
```

B.
```
       6 R4
   5)349
    -35
      4
```

C.
```
       4 R1
   3)121
    -12
      1
```

Tell which problems have errors, and describe the mistakes made.

_____

_____

_____

_____

MORE CALCULATOR MISTAKES
Square Units and Triangular Units

LESSON 64

Name _____

PART A
WARM-UP ACTIVITY: BASIC FACTS AND NEAR EXTENDED FACTS

Activity 1: Find the missing values in the following division facts. R stands for a missing value.

1. 20 ÷ _____ = 4
2. 50 ÷ _____ = 5
3. 20 ÷ _____ = 2
4. 70 ÷ _____ = 10
5. 30 ÷ _____ = 10
6. 60 ÷ _____ = 10

Activity 2: Rewrite the following problems as near extended facts. Do not solve the near extended facts.

1. 31)269

   Near Extended Fact: _____)_____

2. 59)729

   Near Extended Fact: _____)_____

3. 72)487

   Near Extended Fact: _____)_____

PART B
COMMON CALCULATOR ERRORS

Activity: A student entered the following problem on a calculator.

$$21\overline{)192}$$

Here is the answer the student got.

```
6.1428514285
```

1. Estimate what the answer should be by finding a near fact.

2. Find the exact answer on a calculator.

Student Workbook • Number Sense · Unit 4 Division · 167

3. Describe the error the student made.

_____

_____

_____

_____

PART C
FINDING SQUARE AND TRIANGULAR UNITS

Activity: The design below is going to be used on the floor at the entrance to the Arena Building. Look at the design closely. Some of the shapes in the design are the same. You can figure out the size of the different shapes by counting the number of square units. But you can figure out its size more closely by counting the number of triangular units.

Think carefully about how you will answer the questions that follow the design. Hint: You can find the answers to these questions without counting the units every time.

1. What is the total number of square units in both large triangles? What is the total number of triangular units in both large triangles?

2. What is the total numbers of square units in both small triangles? What is the total number of triangular units in both small triangles?

3. What is the total number of square units in the square shapes? What is the total number of triangular units in the square shapes?

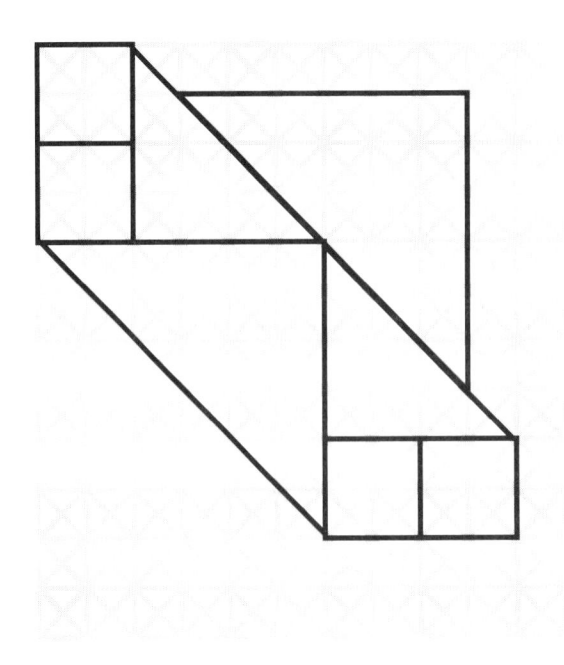

LESSON 65

UNIT 4 REVIEW

Name _____

PART A
BASIC AND EXTENDED FACTS

Activity: Solve the following basic and extended facts.

1. 56 ÷ 8 = _____   560 ÷ 8 = _____   5,600 ÷ 8 = _____
2. 49 ÷ 7 = _____   490 ÷ 7 = _____   4,900 ÷ 7 = _____
3. 32 ÷ 4 = _____   320 ÷ 4 = _____   3,200 ÷ 4 = _____
4. 27 ÷ 9 = _____   270 ÷ 9 = _____   2,700 ÷ 9 = _____

PART B
FACT FAMILIES

Activity: Write fact families for the following groups of numbers. Fill in the missing values for X, Y, and Z.

1. X, 7, and 63   2. 6, 8, and Y   3. 5, Z, and 35

_____   _____   _____

_____   _____   _____

_____   _____   _____

_____   _____   _____

Student Workbook • *Number Sense*   Unit 4 Division

PART C
DIVISION AND NUMBER LINES

Activity: Solve the following problems and show how you got your answers on the number lines. Use a scale that fits the problem.

1. $7\overline{)21}$

⟵――――――――――――――――⟶

2. $4\overline{)16}$

⟵――――――――――――――――⟶

3. $80\overline{)160}$

⟵――――――――――――――――⟶

PART D
NEAR FACTS

Activity: Estimate the following problems by finding the closest basic fact (near fact).

1. $6\overline{)43}$

   Near Fact: $\overline{)}$

2. $7\overline{)50}$

   Near Fact: $\overline{)}$

PART E
LONG DIVISION

Activity: Solve the following problems using traditional long division.

1. 6)372

2. 7)504

PART F
NEAR EXTENDED FACTS

Activity: Rewrite the following problems as near extended facts. Solve the near extended facts.

1. 31)249

   Near Extended Fact:)

2. 39)281

   Near Extended Fact:)

PART G
PULLING OUT 10s AND SOLVING THE BASIC FACT

Activity: Fill in the missing items in the table. You are to rewrite the near extended facts by pulling out the 10s. Then, you are to rewrite the problem as a basic fact and solve the basic fact.

Near Extended Fact	Pull Out the 10s	Basic Fact
60)240	6 × 10)24 × 10	4 6)24
80)720		
50)250		
40)160		

Student Workbook • *Number Sense*   Unit 4 Division • Lesson 65   **171**

PART H
COMMON CALCULATOR ERRORS

Activity: A student entered the following problem on a calculator.

$$8\overline{)59}$$

Here is the answer the student got.

$$.13559322033898$$

1. Estimate what the answer should be by finding a near fact.
2. Find the exact answer on a calculator.
3. Describe the error the student made.

_____

_____

_____

_____

ARRAYS OF NUMBERS 1 TO 15

Review: Basic Operations and Bar Graphs

LESSON 66

Name _____

PART A
WARM-UP ACTIVITY: BASIC FACTS

Activity: Solve the following basic facts. Replace the X, Y, or Z with the correct number.

1. 9 + 4 = ____
2. ____ − 4 = 8
3. ____ − 9 = 7
4. 8 * 6 = ____
5. ____ * 9 = 54
6. ____ ÷ 8 = 6
7. 72 ÷ 9 = ____
8. 7 + ____ = 15
9. ____ * 4 = 16

PART B
ARRAYS FOR 1-15

Activity: Draw arrays for the numbers 1 through 15. Use the grid as your work space. *The arrays for numbers 1 through 5 are done for you as examples*. Notice that the number 4 has two different arrays. As the numbers get larger, more and more will have more than one array. Share your ideas about why this is so as your class creates a large poster of the arrays you have drawn to be displayed in the classroom.

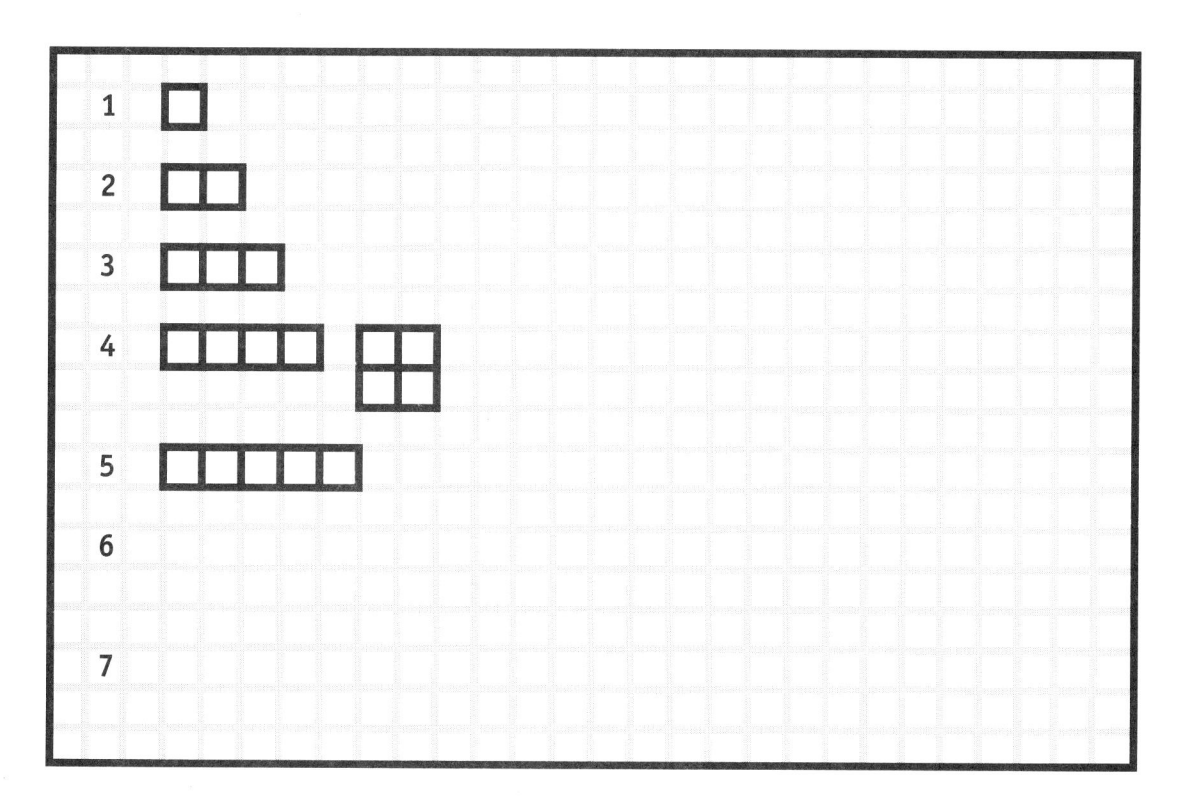

Student Workbook • *Number Sense*   Unit 5 Factors, Patterns, and Multiples   173

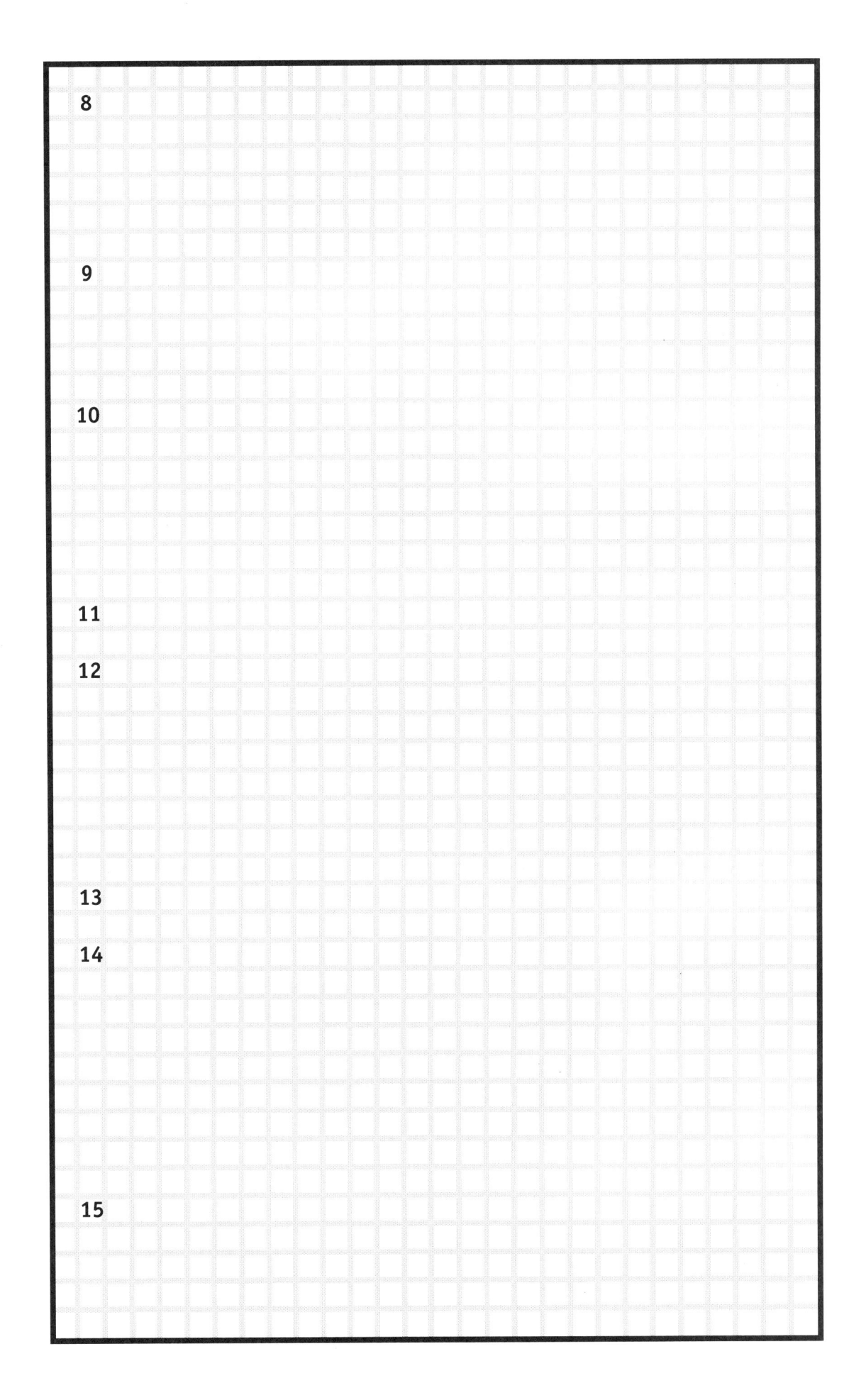

8

9

10

11

12

13

14

15

ARRAYS OF NUMBERS 16 TO 25
Patterns of Numbers

LESSON 67

Name _____

PART A
WARM-UP ACTIVITY: FACT REVIEW

Activity: Solve the following facts and extended facts.

1. $5 \times$ ____ $= 45$
2. $60 \times 2 =$ ____
3. $500 + 600 =$ ____
4. $90 \div 3 =$ ____
5. ____ $- 20 = 70$
6. ____ $\times 7 = 63$
7. $400 +$ ____ $= 1{,}200$
8. $54 \div 6 =$ ____
9. $24 \div$ ____ $= 4$

PART B
ARRAYS FOR 16–25

Activity: Draw arrays for the numbers 16 through 25. Use the grid as your work space. Share your ideas about the arrays as your class completes the large poster of arrays to be displayed in your classroom.

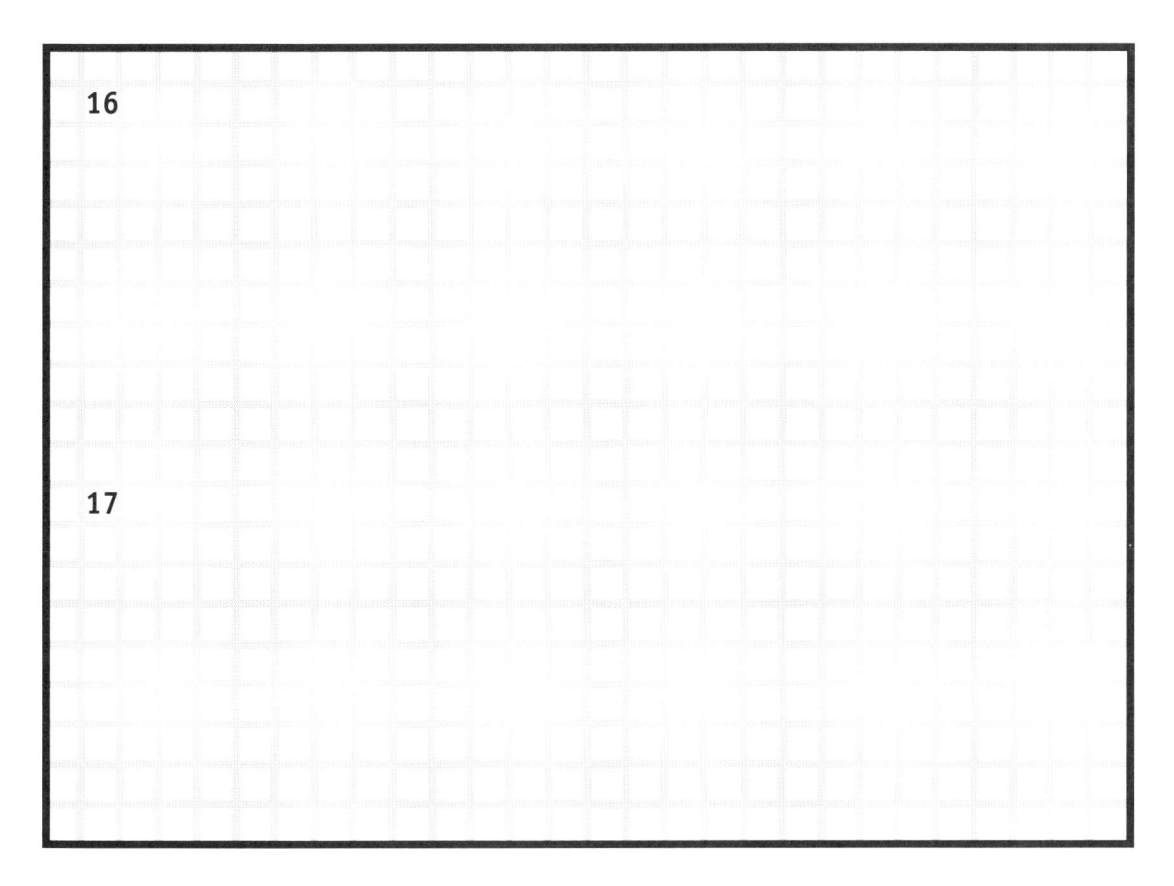

16

17

Student Workbook • Number Sense   Unit 5 Factors, Patterns, and Multiples   175

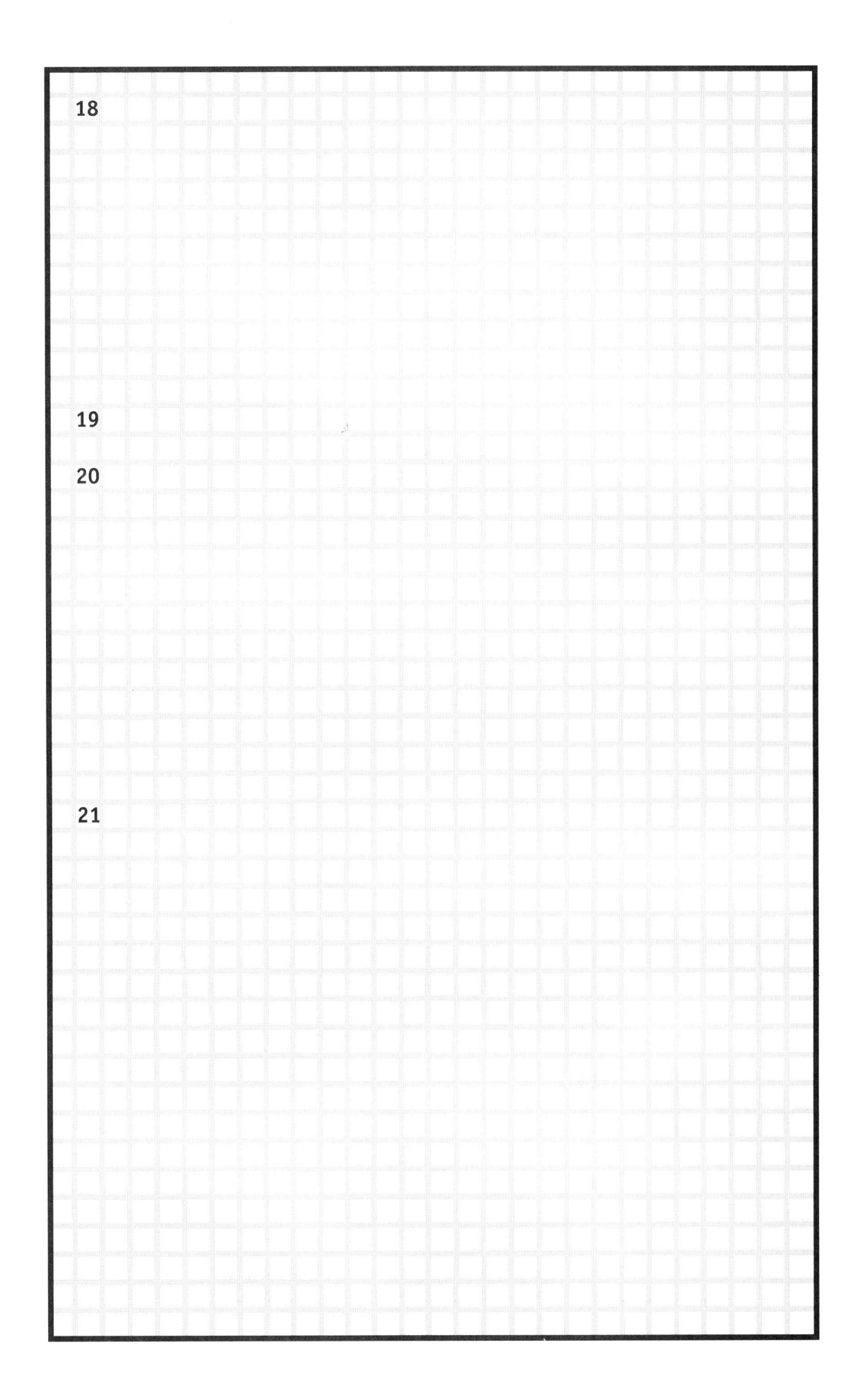

18

19

20

21

22

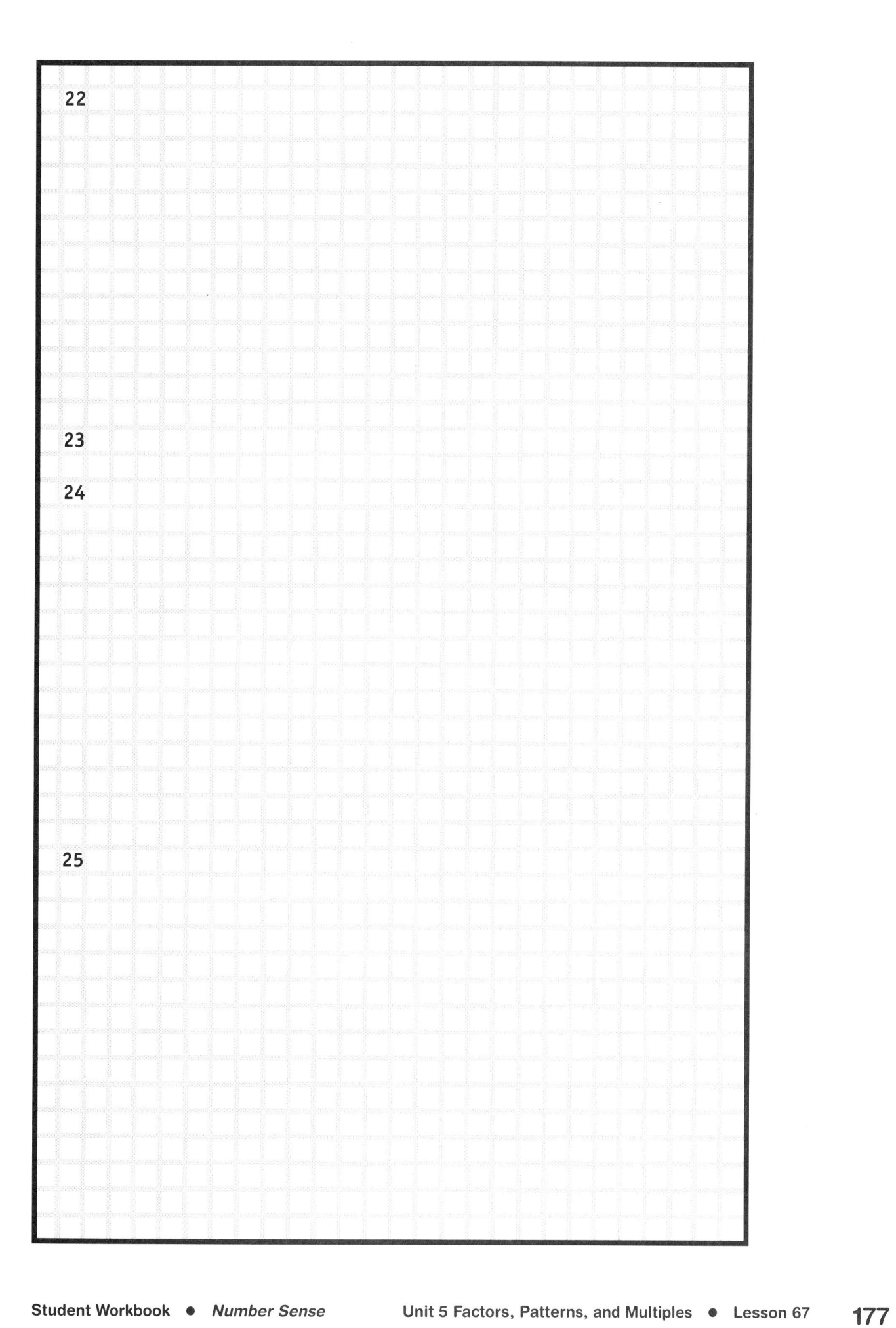

23

24

25

FROM ARRAYS TO FACTORS
Estimating the Size of Irregularly Shaped Objects

LESSON 68

Name _____

PART A
WARM-UP ACTIVITY: BASIC FACTS

Activity: Solve the following basic multiplication facts.

1. 3 × 4 = ____
2. 2 × 6 = ____
3. 2 × 9 = ____
4. 3 × 6 = ____
5. 3 × 8 = ____
6. 4 × 6 = ____

PART B
WRITING THE FACTORS

Activity: Write the multiplication problems *and* the factors for the numbers 1 through 25 in the following table. Some of the work has been done for you. Remember that factors are *not* the multiplication problems, they are the numbers that make up the problems. When writing the factors, be sure that you don't write the same number twice. That means no duplicates!

Number	Multiplication Problems	Factors
1	1 × 1 = 1	1
2	1 × 2 = 2	1, 2
3	1 × 3 = 3	1, 3
4	1 × 4 = 4 2 × 2 = 4	1, 2, 4
5		
6		
7		

Student Workbook • Number Sense   Unit 5 Factors, Patterns, and Multiples   179

Number	Multiplication Problems	Factors
8		
9		
10		
11		
12	$1 \times 12 = 12$ $2 \times 6 = 12$ $3 \times 4 = 12$	1, 2, 3, 4, 6, 12
13		
14		
15	$1 \times 15 = 15$ $3 \times 5 = 15$	1, 3, 5, 15
16		
17		
18		
19		
20		
21		
22		
23		
24		
25		

PART C
FIND THE NUMBER OF SQUARE UNITS USING RECTANGLES

Activity: Use rectangles to estimate the number of square units in each object. You may draw the rectangle inside *or* outside the object to get your estimate. Be sure to multiply to figure out the number of square units.

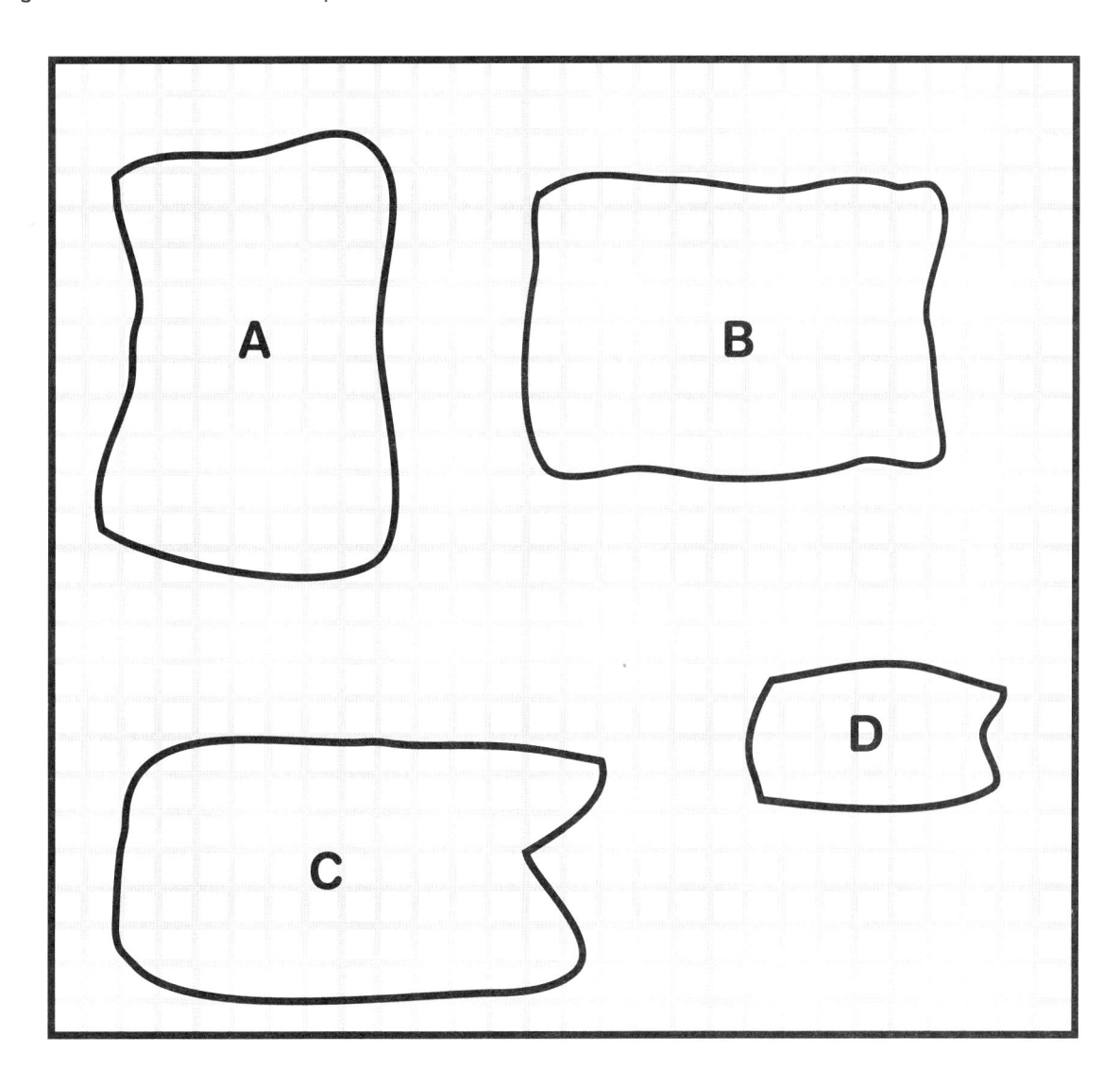

FACTOR RAINBOWS
Areas of Rectangles and Squares

Name _____

LESSON 69

PART A
WARM-UP ACTIVITY: MULTIPLICATION FACTS AND PROBLEMS

Activity: Write all the multiplication facts and multiplication problems for the following numbers. The first one is written in a way to help you get started. Remember, no duplicates!

1. 24 → 1 × ____ = 24
2 × ____ = 24
____ × 8 = 24
4 × ____ = 24

2. 16 → _____
_____
_____

3. 15 → _____
_____

4. 18 → _____
_____
_____

PART B
DRAWING FACTOR RAINBOWS TO FIND FACTORS

Activity: Draw factor rainbows for the following numbers, then list all the factors.

Draw the factor rainbow for the number 21.   **EXAMPLE**

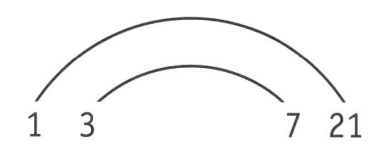

1 3   7 21

The factors of 21 are 1, 3, 7, and 21.

Student Workbook • *Number Sense*   Unit 5 Factors, Patterns, and Multiples   **183**

1. Draw the factor rainbow for the number 16.

   The factors of 16 are: _____

2. Draw the factor rainbow for the number 15.

   The factors of 15 are: _____

3. Draw the factor rainbow for the number 12.

   The factors of 12 are: _____

4. Draw the factor rainbow for the number 7.

   The factors of 7 are: _____

PART C
FINDING AREA OF RECTANGLES AND SQUARES

Activity: In the grid, draw four more rectangular objects. They can be rectangles or squares. Label them B, C, D, and E. Next to each rectangular object, write the area. Look at the example to see how you should write the areas.

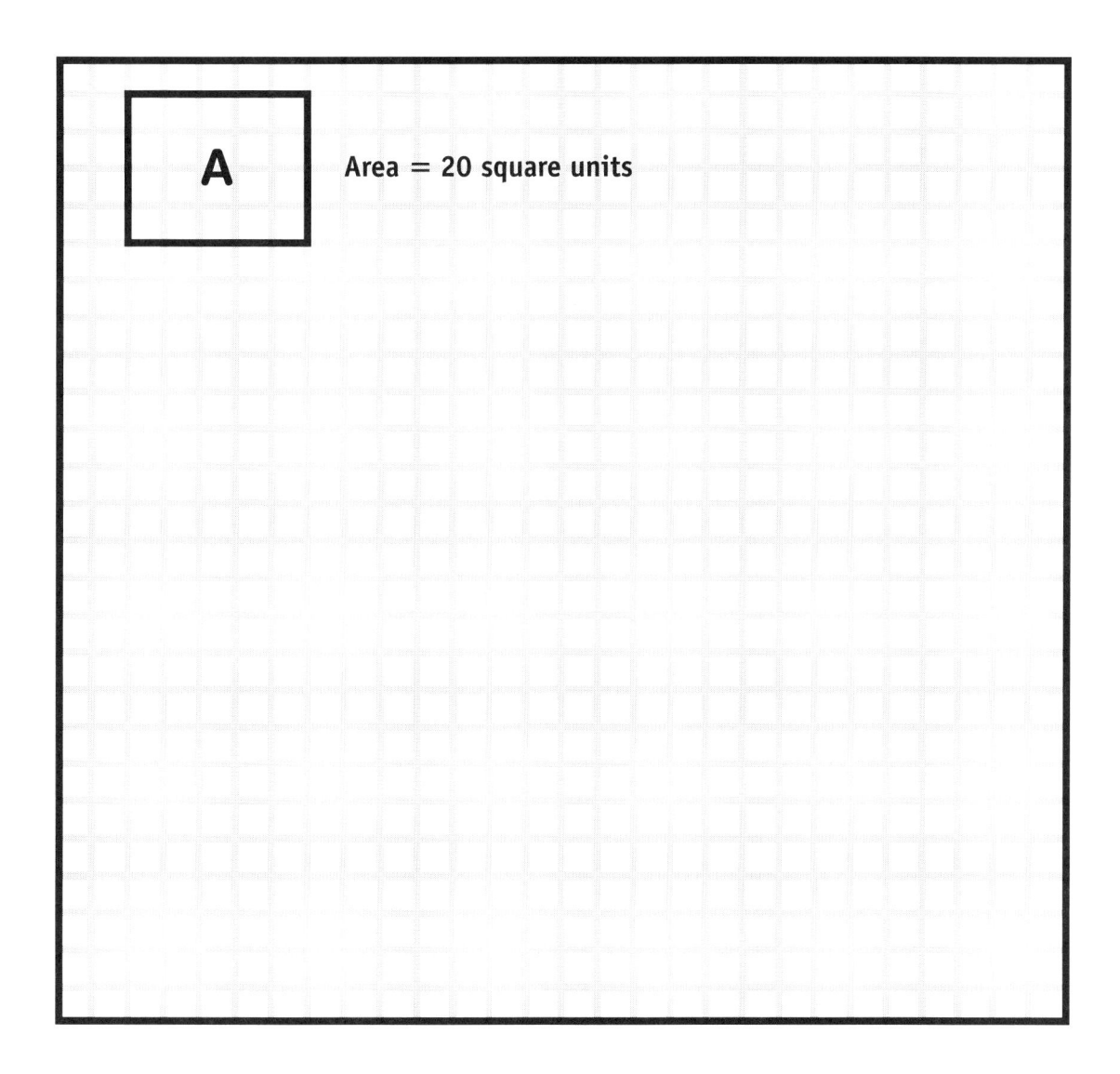

A   Area = 20 square units

WRITING FACTOR LISTS
Reviewing Whole Number Operations

LESSON 70

Name _____

PART A
WARM-UP ACTIVITY: FINDING FACTORS

Activity: For each problem, a number is given followed by a list of numbers. Circle the numbers in the list that are factors of the first number.

16 | 1 | 2 | 3 | 4 | 5 | 6 | 8 | 9 | 16   **EXAMPLE**

You would circle the following:

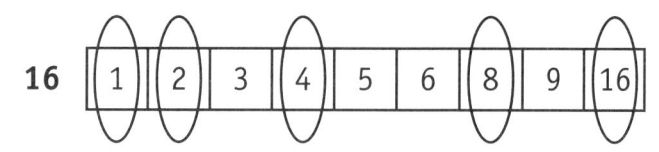

16 | (1) | (2) | 3 | (4) | 5 | 6 | (8) | 9 | (16)

1. 9 | 1 | 2 | 3 | 4 | 5 | 7 | 9 | 11 | 18 | 27 | 36 | 45

2. 12 | 1 | 2 | 3 | 4 | 5 | 6 | 7 | 11 | 12 | 13 | 15

3. 15 | 1 | 3 | 5 | 7 | 9 | 11 | 13 | 15 | 17 | 19

4. 7 | 1 | 2 | 3 | 4 | 5 | 6 | 7 | 8 | 9 | 10

5. 21 | 1 | 2 | 3 | 4 | 5 | 7 | 9 | 10 | 11 | 13 | 21 | 24 | 25

Student Workbook • *Number Sense*   Unit 5 Factors, Patterns, and Multiples   **187**

PART B
MORE PRACTICE WITH FACTOR RAINBOWS

Activity: Draw factor rainbows for the following numbers. Also write the factor lists for the numbers.

1. 25

   The factor list for 25 is: _____

2. 27

   The factor list for 27 is: _____

3. 32

   The factor list for 32 is: _____

4. 36

   The factor list for 36 is: _____

COMPOSITE AND PRIME NUMBERS
Perimeter

LESSON 71

Name _____

PART A
WARM-UP ACTIVITY: FINDING FACTORS

Activity: Circle the numbers in the following lists that are factors for the numbers indicated.

1. 27:   1   2   3   5   9   21   22   27   30

2. 32:   1   2   3   4   8   11   15   16   22   23   31   32   34

3. 40:   1   2   3   4   5   8   10   15   20   25   30   40   45   50

4. 9:   1   2   3   9   18   27   36   45

PART B
NUMBERS WITH ONLY ONE ARRAY

Activity 1: Look at your classroom poster of arrays for the numbers 1–25. Find all the numbers that have only one array (do not include the number 1—it is a special case). These numbers are the *prime numbers*.

Circle the prime numbers between 2 and 25 in the following number grid. Remember, the number 1 is a special case. It is *not* prime.

1	2	3	4	5	6	7	8	9	10
11	12	13	14	15	16	17	18	19	20
21	22	23	24	25					

List the prime numbers here: _____

Student Workbook • Number Sense   Unit 5 Factors, Patterns, and Multiples   189

Activity 2: Draw factor rainbows for the following prime numbers and write their factor lists.

1. Draw the factor rainbow for the number 17.

   The factor list for 17 is: _____

2. Draw the factor rainbow for the number 23.

   The factor list for 23 is: _____

3. Draw the factor rainbow for the number 31.

   The factor list for 31 is: _____

4. Write a sentence that describes what prime numbers are.

   _____
   _____
   _____

PART C
JOSÉ'S GARDENS

Activity: José has five different vegetables planted in a garden in his backyard. Each is planted in a different part of the garden. Suppose that José wants to put fencing around each of the different vegetable patches. Answer the following questions about perimeter. For this activity, assume that each square on the grid has a base of 1 meter and a height of 1 meter.

1. How much fencing will he need for each vegetable patch (that is, what is the perimeter of each of the vegetable patches)?

2. How much bigger is the perimeter for the corn patch than the perimeter for the tomato patch?

3. José is thinking about getting rid of his carrot and pea patches. He wants to make his corn patch larger. It will still have a base of 18 meters, but it will now have a height of 8 meters. What will the perimeter for this new corn patch be?

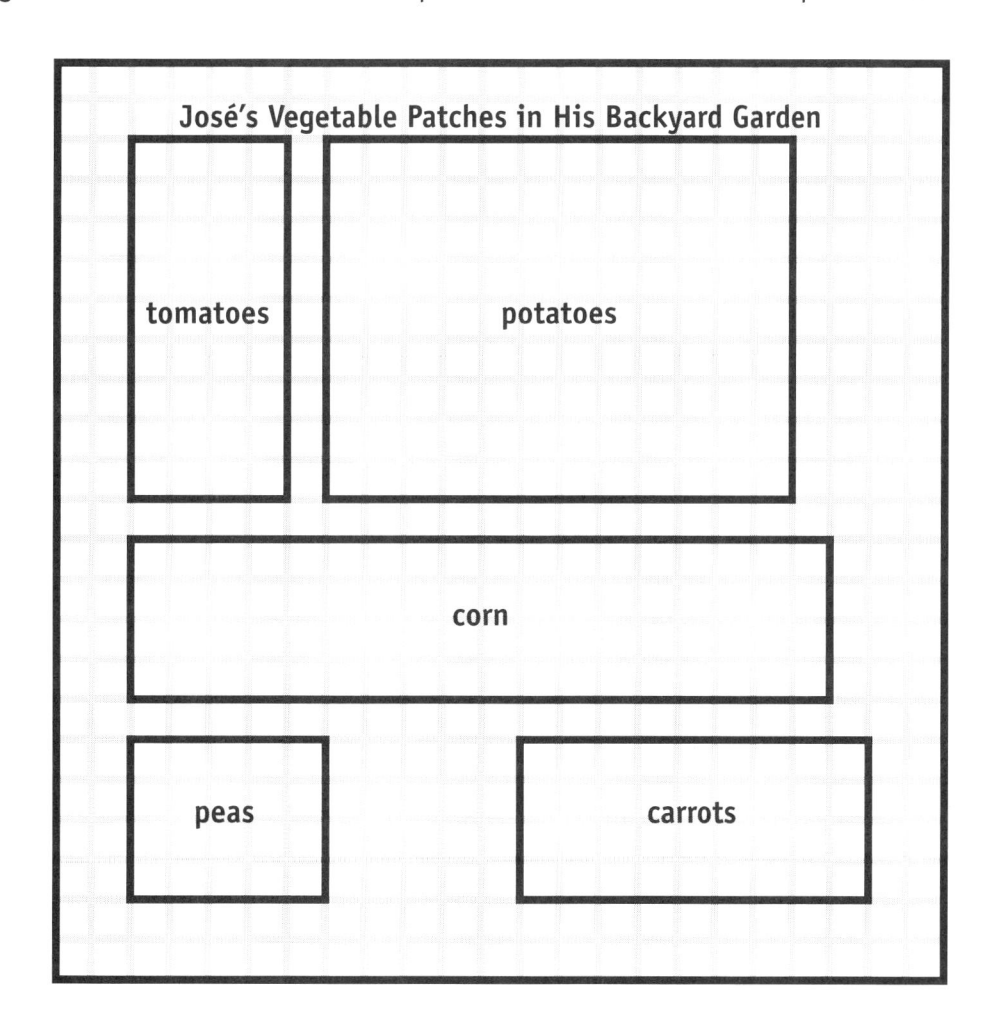

José's Vegetable Patches in His Backyard Garden

MORE ON PRIMES
Numbers Between 1 to 100

LESSON 72

Name _____

PART A
WARM-UP ACTIVITY: EXTENDED FACTS

Activity: Solve the following extended facts. Replace the X, Y, or Z with the correct number.

1. 30 + 60 = ____
2. ____ − 40 = 30
3. ____ − 500 = 200
4. 80 * 5 = ____
5. ____ * 9 = 540
6. ____ ÷ 8 = 100
7. 720 ÷ 9 = ____
8. 70 + ____ = 150
9. ____ * 4 = 160

PART B
EXPLORING PRIMES

Activity 1: Look at the prime numbers between 1 and 100. They are the shaded numbers in the following grid.

Primes Numbers Between 1 and 100

1	2	3	4	5	6	7	8	9	10
11	12	13	14	15	16	17	18	19	20
21	22	23	24	25	26	27	28	29	30
31	32	33	34	35	36	37	38	39	40
41	42	43	44	45	46	47	48	49	50
51	52	53	54	55	56	57	58	59	60
61	62	63	64	65	66	67	68	69	70
71	72	73	74	75	76	77	78	79	80
81	82	83	84	85	86	87	88	89	90
91	92	93	94	95	96	97	98	99	100

List two things you notice about these prime numbers:

1. _____
2. _____

Student Workbook • *Number Sense*   Unit 5 Factors, Patterns, and Multiples   **193**

Activity 2: The following number grid shows all the whole numbers between 501 and 600.

501	502	503	504	505	506	507	508	509	510
511	512	513	514	515	516	517	518	519	520
521	522	523	524	525	526	527	528	529	530
531	532	533	534	535	536	537	538	539	540
541	542	543	544	545	546	547	548	549	550
551	552	553	554	555	556	557	558	559	560
561	562	563	564	565	566	567	568	569	570
571	572	573	574	575	576	577	578	579	580
581	582	583	584	585	586	587	588	589	590
591	592	593	594	595	596	597	598	599	600

There are 14 prime numbers between 501 and 600. Let's see if you can guess what some of them are. It would take a lot of lucky guesses to find all 14. However, you can make the task easier by crossing out numbers that you know are not prime.

Are there any numbers (or groups of numbers) that you can cross out right away—numbers that could not possibly be prime numbers? *Use your pencil and cross out those numbers.*

Write your reasons for crossing out these numbers (or for *not* crossing out other numbers):

_____

_____

Now make a guess about which of the remaining numbers are prime. List five of them.

_____ _____ _____ _____ _____

Be ready to share your guesses in class and tell why you think those numbers are prime.

As your teacher reads the actual list, circle the prime numbers in the grid.

How many of your guesses were correct? _____

AREA
Different Shapes

Name _____

LESSON 73

PART A
WARM-UP ACTIVITY:
BASIC AND EXTENDED FACTS WITH VARIABLES

Activity 1: Solve the following extended facts.

1. 30 × ____ = 150
2. 600 × 2 = ____
3. 60 + 80 = ____
4. 120 ÷ 4 = ____
5. ____ − 50 = 100
6. ____ × 7 = 630
7. 700 + ____ = 1,200
8. 600 ÷ 6 = ____
9. 240 ÷ ____ = 40

Activity 2: Circle all the prime numbers in the following list.

2   3   5   6   9   11   13   15   22   29   31   35

Activity 3: List all of the factors for the following numbers. The first problem is done for you.

Number	Factors
20	1, 2, 4, 5, 10, 20
12	
9	
13	
25	

Student Workbook • *Number Sense* — Unit 5 Factors, Patterns, and Multiples

PART B
AREA OF DIFFERENT SHAPES

Activity 1: In the following grid are two shapes—a rectangle and a triangle. You know how to find the area of the rectangle. How would you find the area of the triangle? *Don't count the squares. Think of a formula that you could create for the triangle based on what you know about the area of the rectangle.*

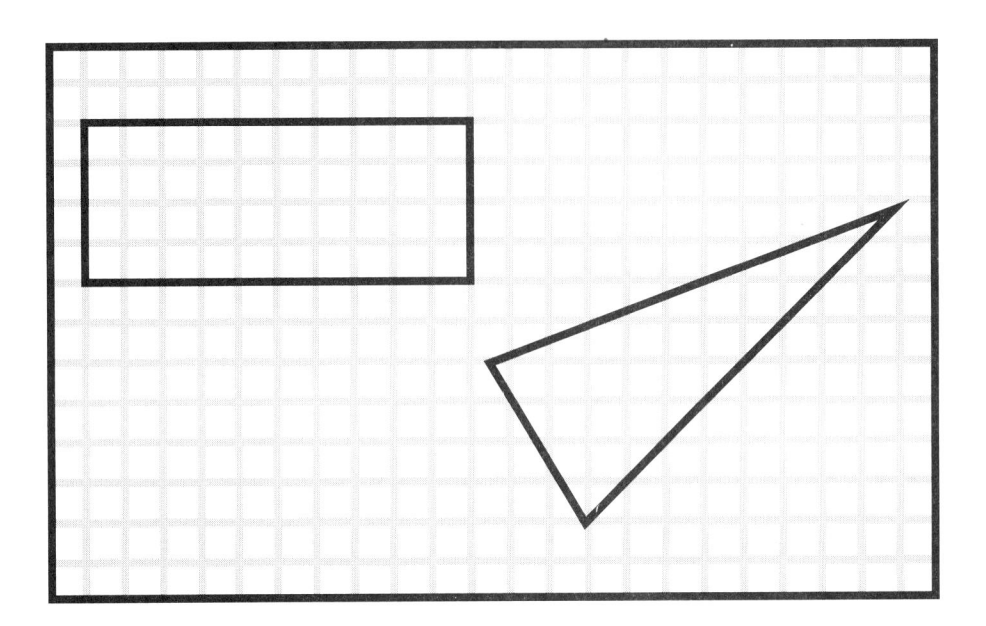

Activity 2: In the following grid are two shapes—a rectangle and a parallelogram. You know how to find the area of the rectangle. How would you find the area of the parallelogram? *Don't count the squares. Think of a formula that you could create for the parallelogram based on what you know about the area of the rectangle.*

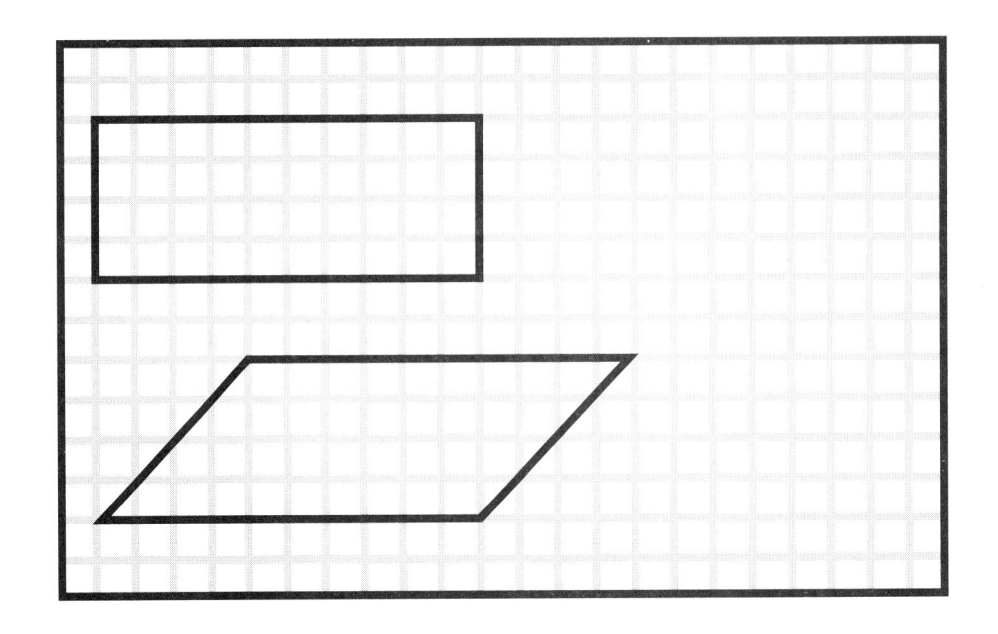

196   Unit 5 Factors, Patterns, and Multiples   •   Lesson 73   Student Workbook   •   *Number Sense*

PRIME FACTOR TREES
Areas and Floor Plans

Name _____

LESSON 74

PART A
WARM-UP ACTIVITY: FACTORS

Activity: List all of the factors for the following numbers. The first problem is done for you.

Number	Factors
10	1, 2, 5, 10
17	
2	
24	
8	
6	

PART B
FINDING PRIMES

Activity 1: Circle the primes in the following list of numbers.

1   2   4   5   8   11   15   21   49

Activity 2: Which numbers in Activity 1 are not prime? List them:

_____

Student Workbook • *Number Sense*   Unit 5 Factors, Patterns, and Multiples   197

Activity 3: Draw prime factor trees for the numbers you listed in Activity 2. Be sure to circle the prime factors.

PART C
LAYING CARPET

Activity: In Lesson 70, you looked at the blueprint for a summer cabin. In this activity, you will be asked some questions about laying carpet and installing tile on the bottom floor of the cabin. The owners want you to carpet the living room and the kitchen. They want you to put tile on the bathroom floor.

1. How many square units of carpet will you need for the two rooms? _____

2. How many square units of tile will you need for the bathroom? _____

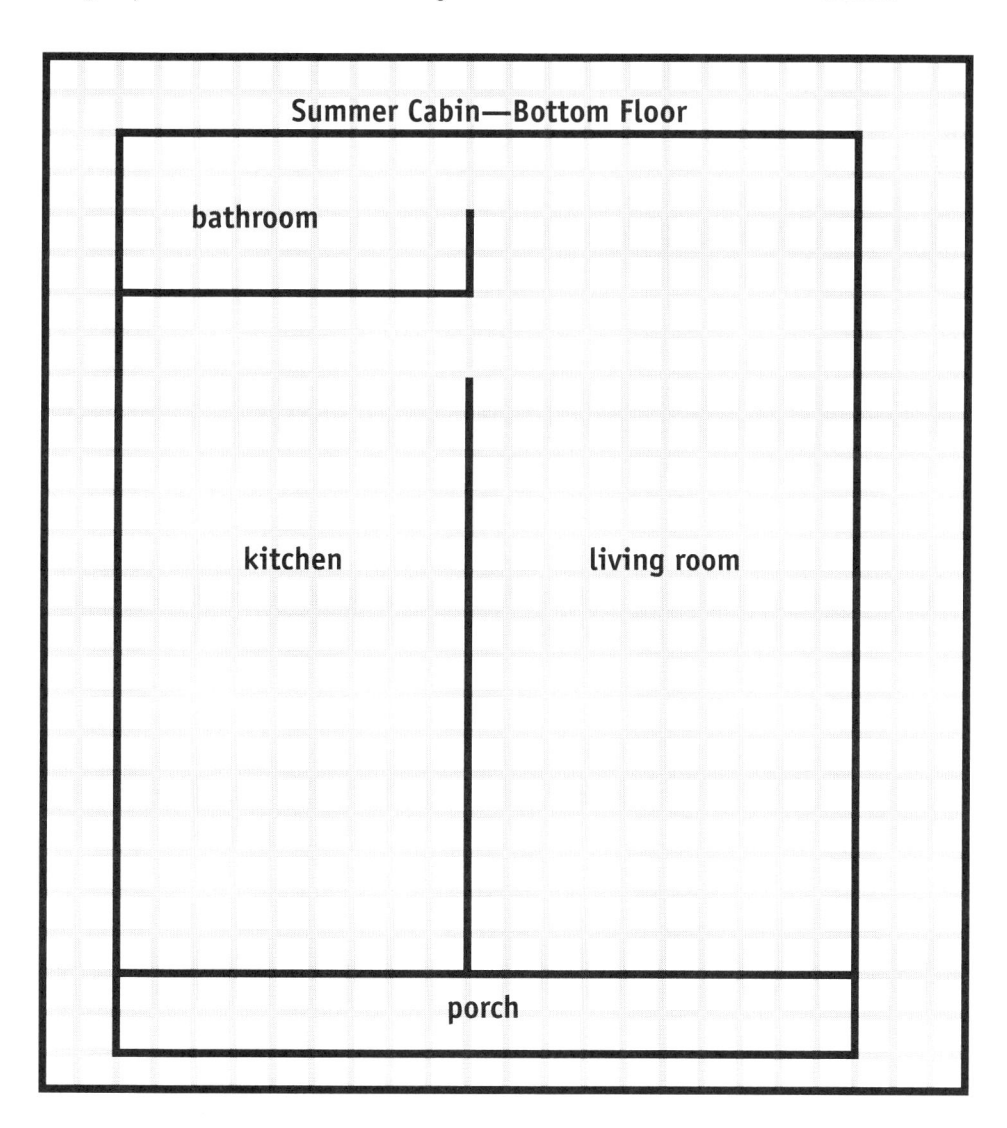

Summer Cabin—Bottom Floor

bathroom

kitchen   living room

porch

MORE PRACTICE ON PRIME FACTOR TREES
Comparing Area and Perimeter

LESSON 75

Name _____

PART A
WARM-UP ACTIVITY: MULTIPLICATION FACTS

Activity 1: Solve the following multiplication facts.

1. 5 × ____ = 45
2. 6 × 7 = ____
3. 7 × 4 = ____
4. ____ × 5 = 20
5. 9 × ____ = 36
6. 7 × 8 = ____

Activity 2: List all of the factors for the following numbers. The first problem is done for you.

Number	Factors
9	1, 3, 9
45	
28	
42	
20	
36	

PART B
PRIME FACTOR TREES

Activity: Draw prime factor trees for the numbers 20 and 28. Be sure to circle the prime factors.

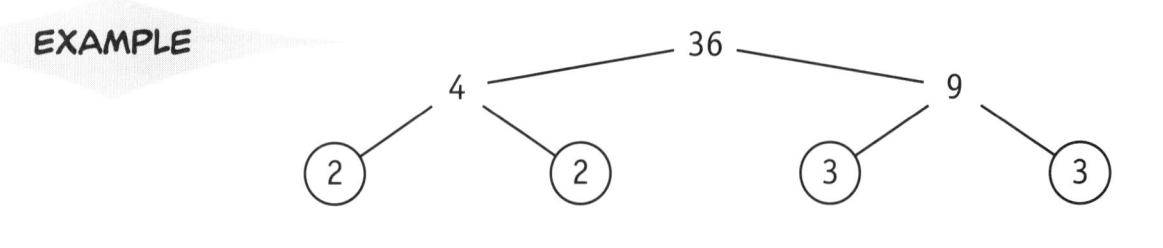

EXAMPLE

```
         36
      4      9
    2   2  3   3
```

20

28

PART C
AREA AND PERIMETER

Activity 1: Draw three rectangles of different sizes on the grid. Find the area and perimeter of each rectangle, and write the values next to the rectangles. Assume that each square on the grid has a height of 1 cm and a base of 1 cm. Then answer the question that follows the grid.

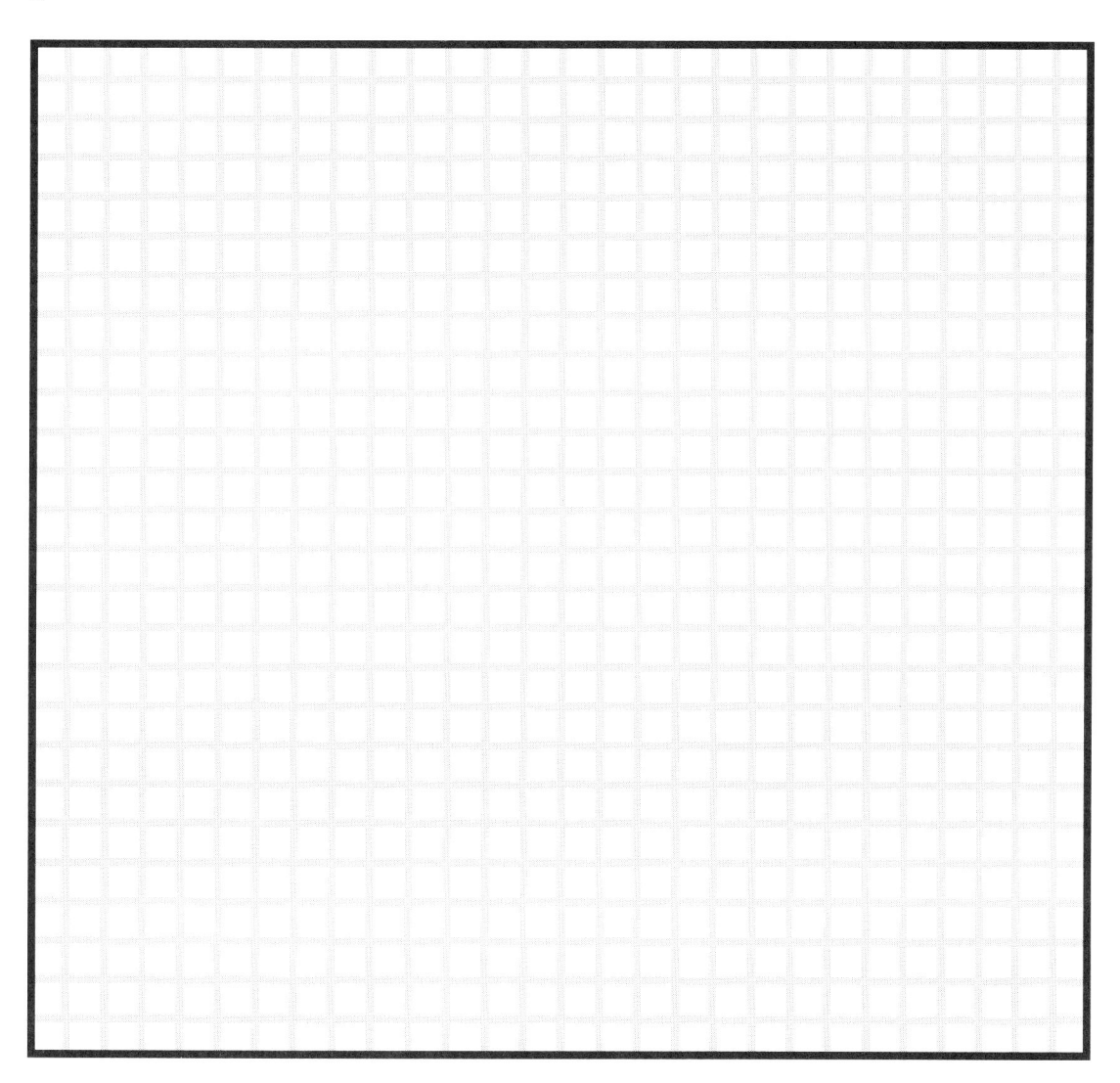

Activity 2: Use your work in Activity 1 to answer this question:

Is the area of a rectangle always bigger than the perimeter?

_____
_____
_____

AREA AND PERIMETER
Which Is Bigger?

Name _____

LESSON 76

PART A
WARM-UP ACTIVITY: PRIME FACTORIZATION

Activity: Fill in the circles for these prime factor trees with the correct numbers. Then complete the prime factorizations.

1.

    21

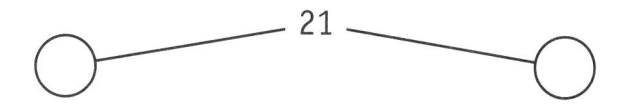

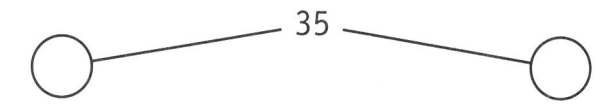

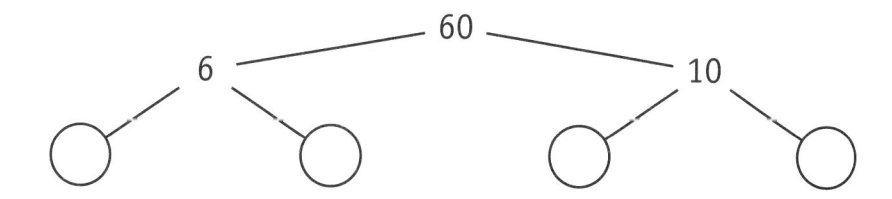

    The prime factorization of 21 is ____ × ____.

2.

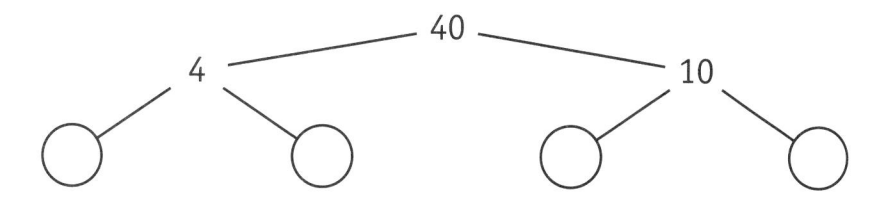

    40
    4   10

    The prime factorization of 40 is ____ × ____ × ____ × ____.

3.

    35

    The prime factorization of 35 is ____ × ____.

4.

    60
    6   10

    The prime factorization of 60 is ____ × ____ × ____ × ____.

Student Workbook • Number Sense   Unit 5 Factors, Patterns, and Multiples   **205**

5.

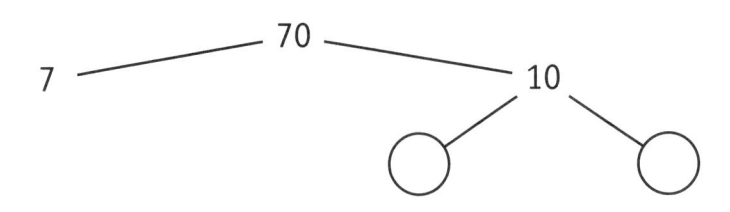

The prime factorization of 70 is ____ × ____ × ____ × ____.

6.

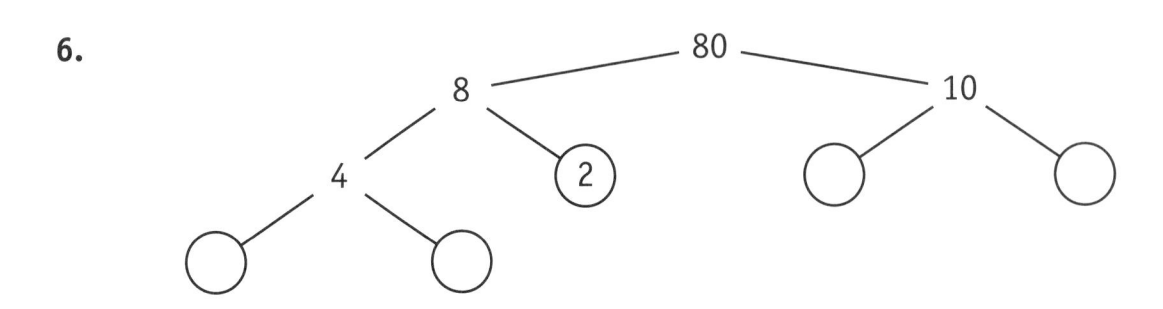

The prime factorization of 80 is ____ × ____ × ____ × ____ × ____.

PART B
AREA AND PERIMETER

A builder wants to build a room that has a *perimeter* of 40 meters. Each square on the following grid is drawn to scale, with each square *representing* a height of 1 meter and a base of 1 meter. (We know the grid is drawn to scale and not exact because a meter is much bigger than these squares. Remember, a meter is about as long as a baseball bat.) Draw as many rectangles as you can on the grid that have perimeters of 40 meters. We have given you two pages of graph paper on which to do this. Also, calculate the area for each rectangle. On the inside of each rectangle, write A = ____, giving the area. *Note: The perimeter should always be 40.*

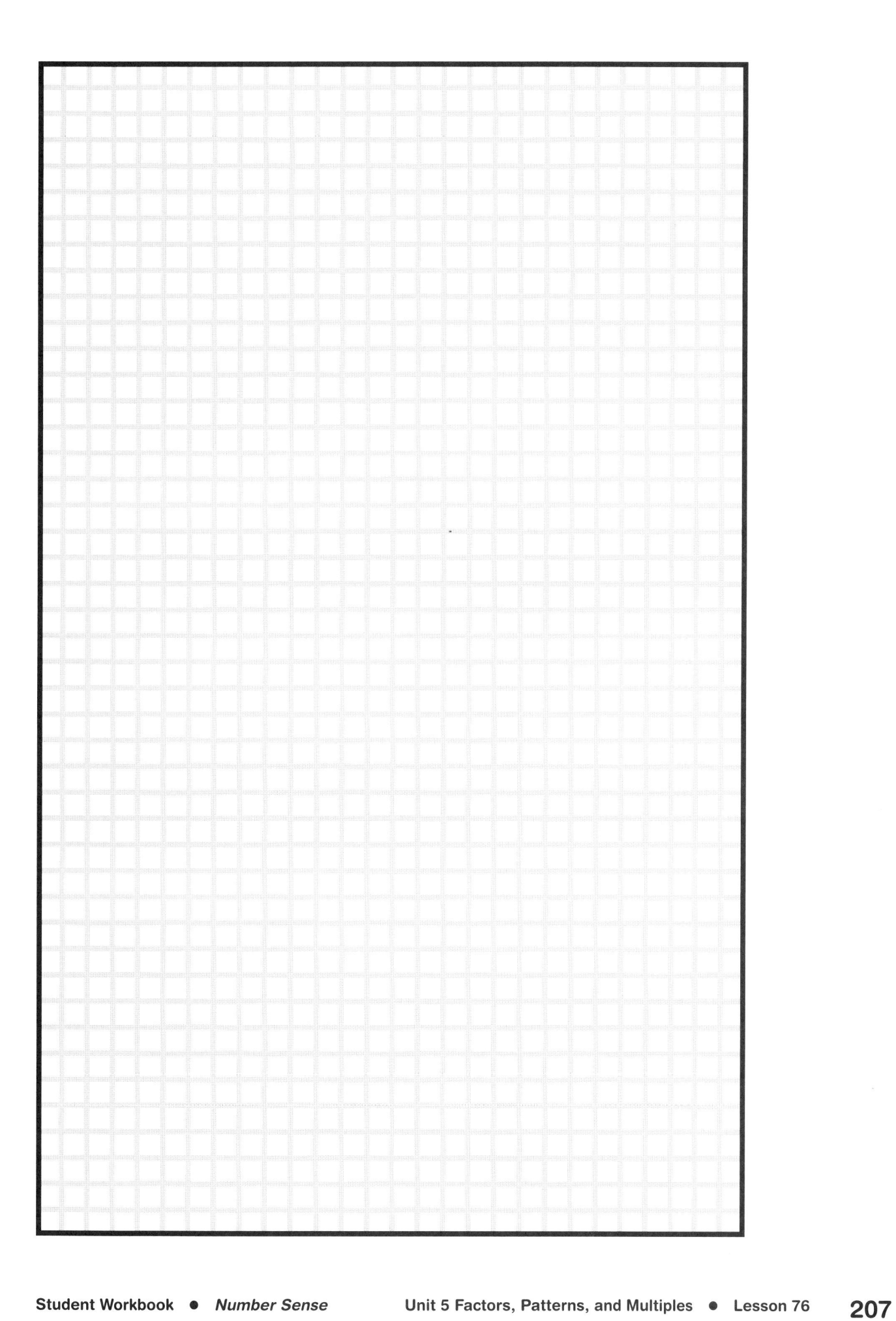

LOOKING FOR PATTERNS IN AREA AND PERIMETER
Comparing Both Measures

LESSON 77

Name _____

PART A
WARM-UP ACTIVITY: MULTIPLICATION FACTS AND EXTENDED FACTS

Activity: Solve the following basic and extended multiplication facts.

1. $7 \times 6 =$ ____
   $6 \times 7 =$ ____
   $7 \times 60 =$ ____
   $7 \times 600 =$ ____

2. $4 \times 5 =$ ____
   $5 \times 4 =$ ____
   $5 \times 40 =$ ____
   $5 \times 400 =$ ____

3. $7 \times 3 =$ ____
   $3 \times 7 =$ ____
   $7 \times 30 =$ ____
   $7 \times 300 =$ ____

PART B
RELATIONSHIP BETWEEN AREA AND PERIMETER

Activity 1: In the last lesson, you drew a lot of rectangles that had different areas and the same perimeter. That perimeter was 40 meters. Today, you are going to see how area and perimeter relate.

First, make an *organized list* that contains the base, height, difference between base and height, perimeter, and area of 10 rectangles. The first should have a base of 19 meters and a height of 1 meter. The second should have a base of 18 meters and a height of 2 meters. Continue in this way up until the last rectangle, which would have a base of 10 meters and a height of 10 meters. We have completed the list for the first and last rectangle for you. You can use your calculator as you fill in the list.

Rectangle	Base	Height	Difference Between Base and Height	Perimeter	Area
1	19	1	18	40	19
2					
3					
4					
5					
6					
7					
8					
9					
10	10	10	0	40	100

Activity 2: Now that you have completed your list, you are to make two bar graphs. Graph A should show the difference between the base and the height for the 10 rectangles. Graph B should show the areas of the rectangles. Make Graph A first. Use the numbers from the "Difference Between Base and Height" column in your table to make this graph.

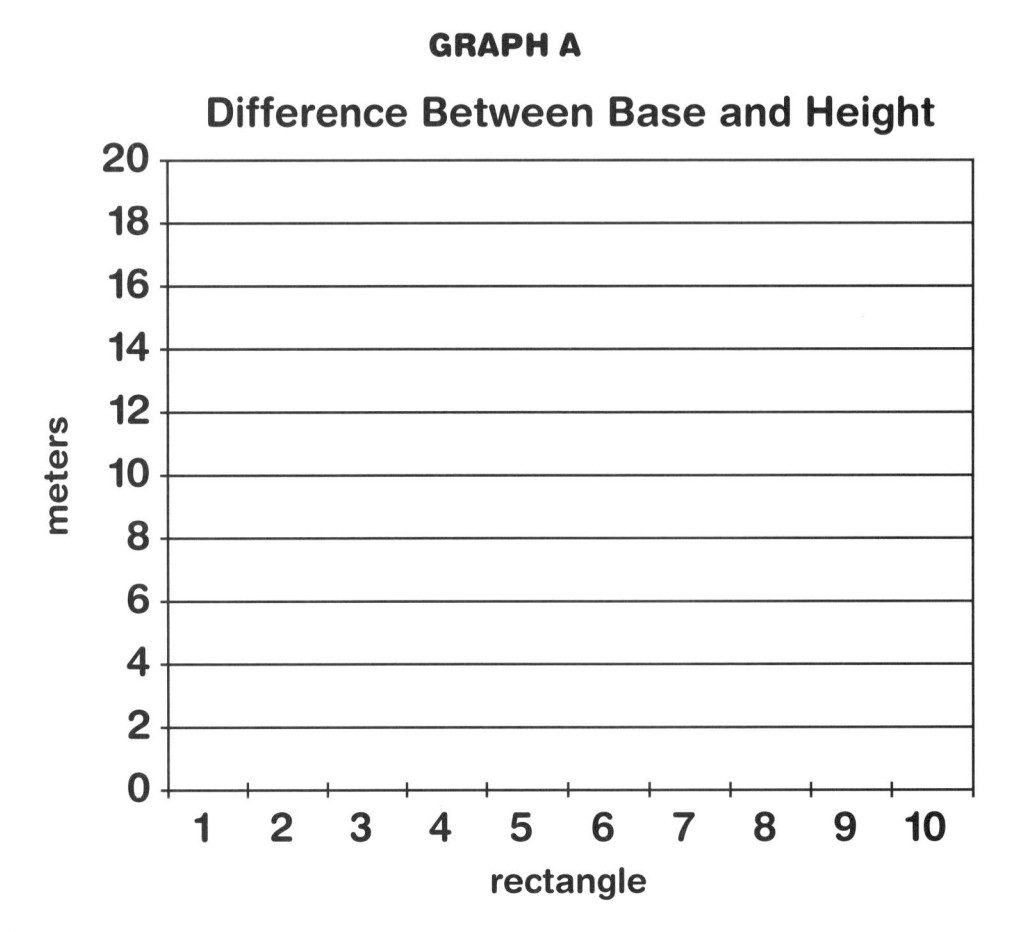

GRAPH A

Difference Between Base and Height

Now make Graph B, showing the area of each rectangle in the organized list. Be sure to approximate your numbers on the graph. They don't have to be exact.

GRAPH B

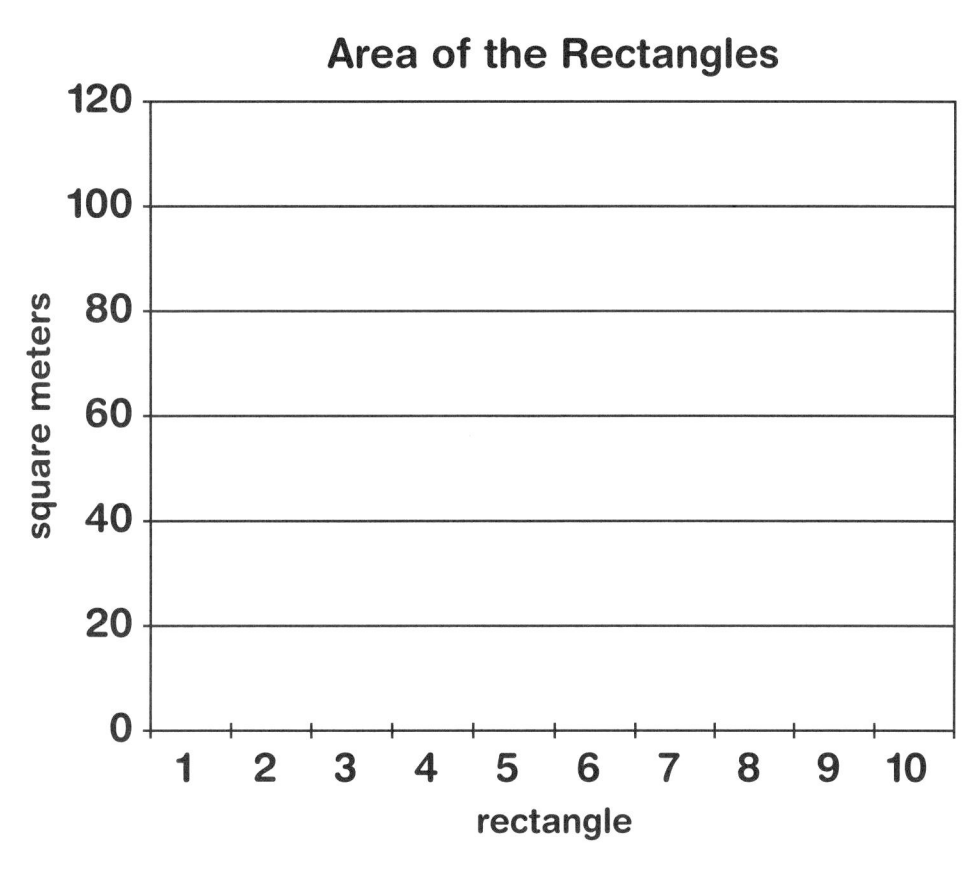

Area of the Rectangles

[Blank graph with y-axis labeled "square meters" from 0 to 120 in increments of 20, and x-axis labeled "rectangle" from 1 to 10.]

Now look at the two graphs and answer these questions.

1. What happens to the area when the difference between the base and height gets smaller?

   _____

2. Of all of the rectangles, what shape has the largest area?

   _____

REVIEW OF LESSONS 66 TO 77

Name _____

LESSON 78

PART A
ARRAYS

Activity: On the grids provided, draw all the arrays for each of the following numbers. Remember, no duplicates!

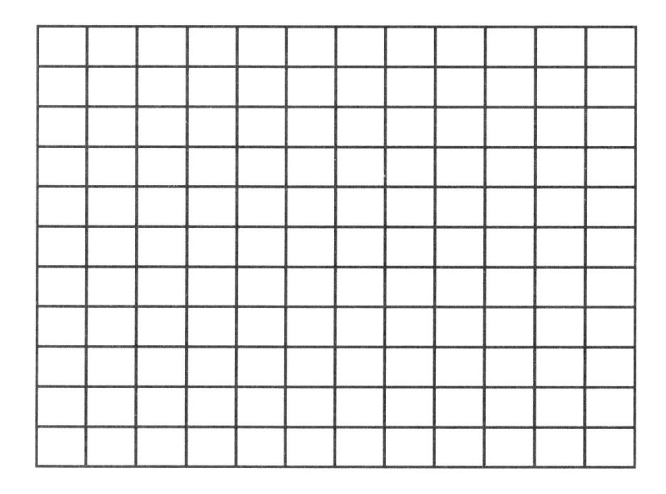

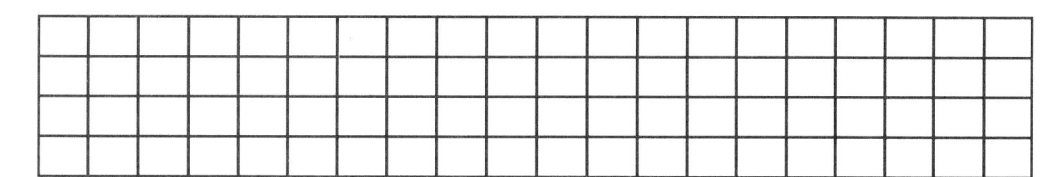

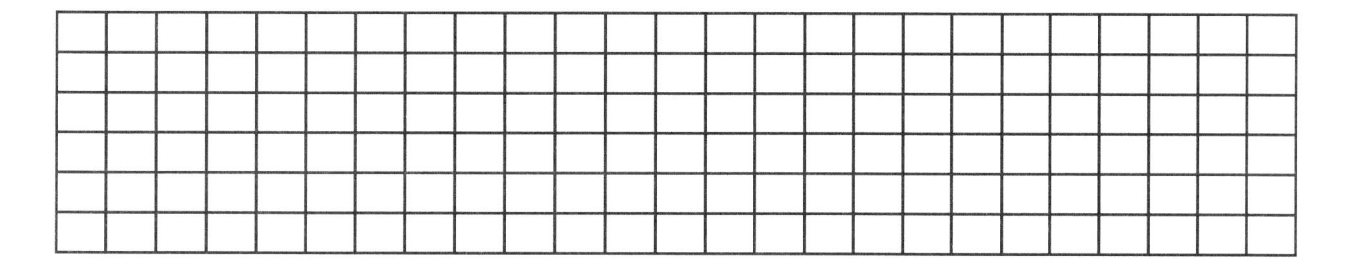

12

17

25

PART B
FACTORS

Activity: For each problem, a number is given followed by a list of numbers. Circle the numbers in the list that are factors of the first number.

1. 16 | 1 | 2 | 3 | 4 | 5 | 6 | 8 | 9 | 16 |

2. 9 | 1 | 2 | 3 | 4 | 5 | 7 | 9 | 11 | 18 |

3. 21 | 1 | 2 | 3 | 4 | 5 | 7 | 9 | 11 | 21 |

4. 31 | 1 | 2 | 3 | 24 | 25 | 27 | 29 | 31 |

PART C
FACTOR RAINBOWS

Activity: Draw the factor rainbows for the following numbers, then list all the factors.

1. 12

2. 15

3. 17

4. 25

PART D
AREA AND PERIMETER OF RECTANGLES AND SQUARES

Activity: Find the area and perimeter of the following rectangles and/or squares with the given dimensions. Remember, area is measured in square units and perimeter in units.

1. Base = 2 cm, Height = 4 cm   Area = _____   Perimeter = _____
2. Base = 3 in., Height = 3 in.   Area = _____   Perimeter = _____
3. Base = 4 m, Height = 10 m   Area = _____   Perimeter = _____

PART E
COMPOSITE AND PRIME NUMBERS

Activity: Tell whether the following numbers are prime or composite by circling the correct response.

1. 12   Prime   or   Composite   5. 23   Prime   or   Composite
2. 15   Prime   or   Composite   6. 29   Prime   or   Composite
3. 17   Prime   or   Composite   7. 33   Prime   or   Composite
4. 21   Prime   or   Composite

PART F
AREAS OF SHAPES

Activity: Find the area of the following shapes. Remember, area is measured in square units.

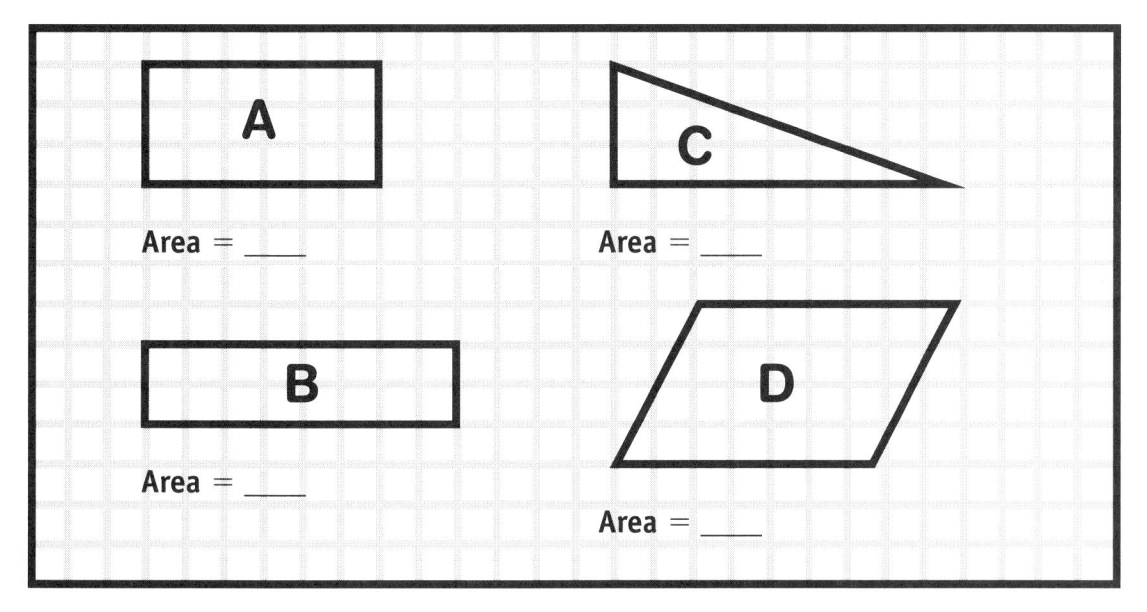

A   Area = _____

B   Area = _____

C   Area = _____

D   Area = _____

PART G
PRIME FACTORIZATION

Activity: Fill in the missing information in the following prime factor trees. Then complete the prime factorizations.

1.

   36 branches to 6 and 6; each 6 branches to two empty circles.

   The prime factorization of 36 is ____ × ____ × ____ × ____.

2.

   50 branches to 2 and 25; 25 branches to two empty circles.

   The prime factorization of 50 is ____ × ____ × ____.

3.

   54 branches to 6 and 9; each branches to two empty circles.

   The prime factorization of 54 is ____ × ____ × ____ × ____.

DIVIDING RULES
Measuring Area

LESSON 79

Name _____

PART A
WARM-UP ACTIVITY: DIVISION FACTS

Activity: Solve the following basic and extended division facts.

1. 10 ÷ 5 ____
2. 150 ÷ 10 = ____
3. 18 ÷ 2 = ____
4. 35 ÷ ____ = 5
5. 10 ÷ 2 = ____
6. 120 ÷ ____ = 10

PART B
USING YOUR DIVIDING RULES

Activity 1: Mark an "X" in the table if the dividing rule works for the number. The dividing rules we are looking at here are divide by 2, divide by 5, and divide by 10. More than one rule may work for a number. Mark *all* rules that work. Then complete Problems 1–3, which follow the table.

Number	Divide by 2	Divide by 5	Divide by 10
10	X	X	X
12			
15			
27			
65			
150			

1. Give an example of a five-digit number that you can divide by 2. ____
2. Give an example of a five-digit number that you can divide by 5. ____
3. Give an example of a five-digit number that you can divide by 10. ____

Student Workbook • *Number Sense*   Unit 5 Factors, Patterns, and Multiples   **217**

Activity 2: Mark an "X" in the table if the dividing rule works for the number. The dividing rules we are looking at here are divide by 3 and divide by 6. More than one rule may work for a number. Mark *all* rules that work. Then complete Problems 1 and 2, which follow the table.

Number	Divide by 3	Divide by 6
6	X	X
12		
15		
21		
24		
51		

1. Give an example of a four-digit number that you can divide by 3. _____
2. Give an example of a four-digit number that you can divide by 6. _____

PART C
MEASURING AREA OF THE DESIGN

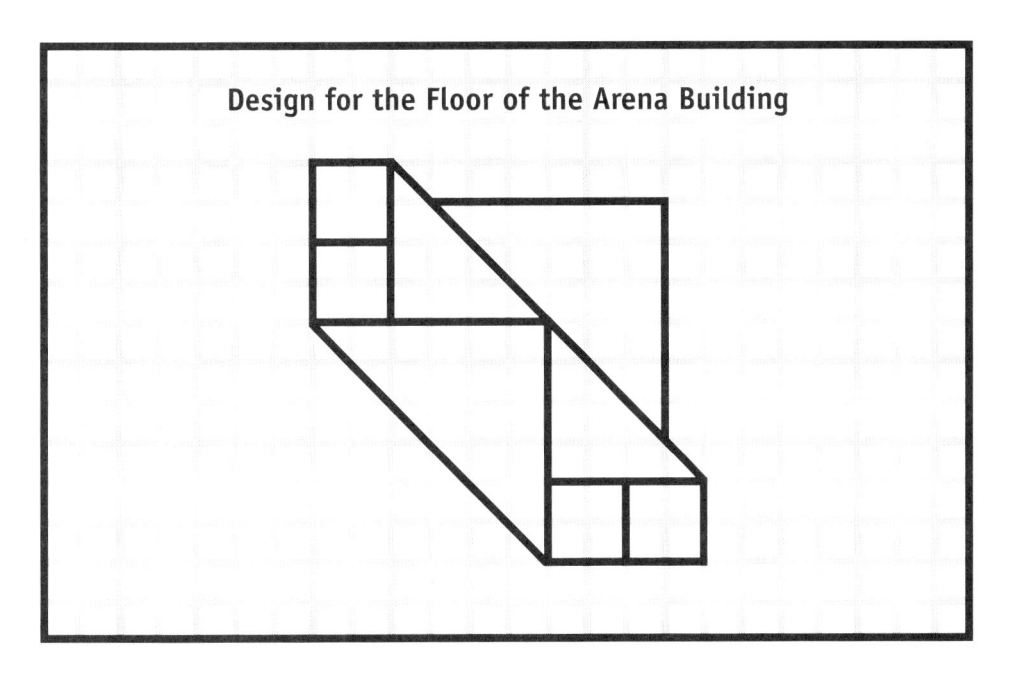

Design for the Floor of the Arena Building

Activity: Answer the following questions about this design. (When we refer to the *design*, we are referring to the part of the drawing within the bold black outline.) Think about how area is calculated using base and height. Be sure to give your answers in square units.

1. What is the area of one of the squares in the design? _____

   What is the total area of all of the squares in the design? _____

2. What is the total area of the two smaller triangles in the design? _____

3. What is the total area of the two larger triangles in the design? _____

4. What is the total area of the entire design (*just* the design)? _____

MORE PRACTICE WITH DIVIDING RULES
Geometry

LESSON 80

Name _____

PART A
WARM-UP ACTIVITY: BASIC AND EXTENDED FACTS

Activity: Solve the following basic and extended facts.

1. 12 ÷ 3 = ____
2. 100 × 10 = ____
3. 30 ÷ ____ = 5
4. 36 ÷ 6 = ____
5. 20 ÷ ____ = 10
6. ____ ÷ 5 = 100
7. 100 ÷ 2 = ____
8. 6 × ____ = 24
9. ____ ÷ 2 = 100

PART B
DIVIDING RULES

Activity: Here are the dividing rules from the last lesson. Use them to fill in the table that follows the list of rules.

2	We can divide a number by 2 if it is an even number.
3	We can divide a number by 3 if we can add up its digits and divide that number by 3 evenly.
5	We can divide a number by 5 if it ends in a 5 or a 0.
6	We can divide a number by 6 if we can divide it by 2 *and* by 3.
10	We can divide a number by 10 if it ends in a 0.

Student Workbook • *Number Sense* — Unit 5 Factors, Patterns, and Multiples

Which rule or rules work for the following numbers? The rules for the first number have been filled in for you.

Number	Rule or Rules
270	2, 3, 5, 6, and 10 rules
555	
78	
393	
76,100	
666	

PART C
COMPARING SHAPES

Activity 1: Look at the these shapes. You are to put them into groups. The shapes in each group should have at least one feature in common. Make between 6 and 10 groups. It is okay to put a shape in a group by itself. On the table that follows, label each group, give the numbers of the shapes in that group, and describe what is the same about the shapes. *Note: You can use each shape only once.* That means that once you use a shape, you can't use it again.

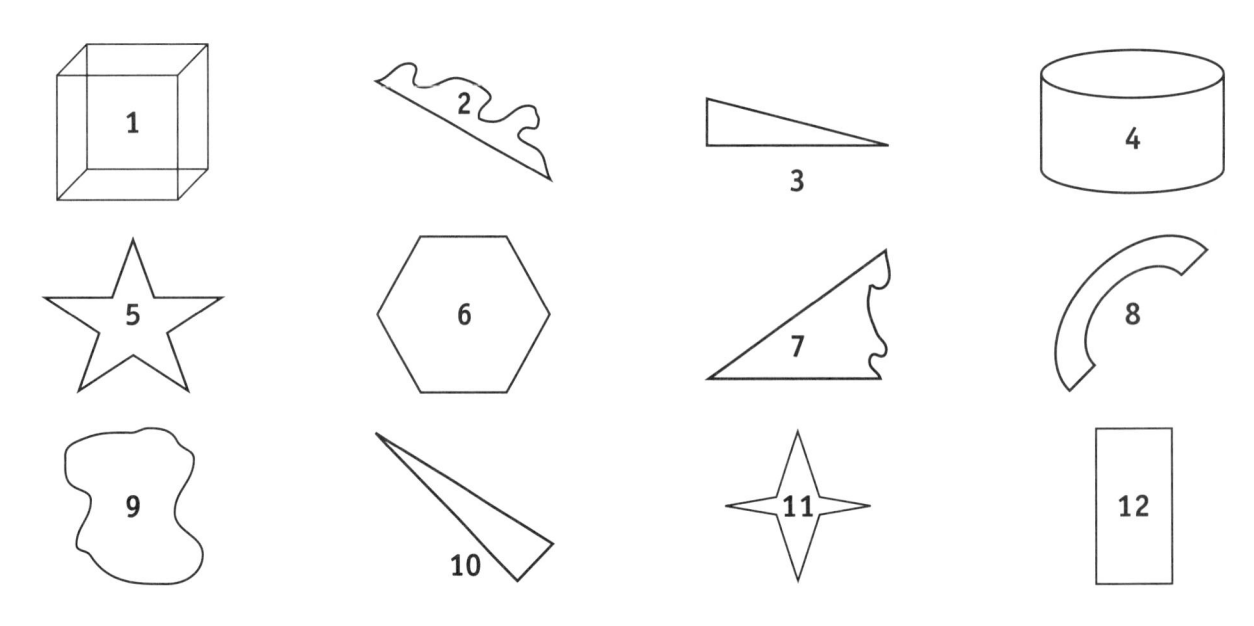

Group	Label for Group	Numbers for the Shapes in That Group	Describe What Is the Same About the Shapes in This Group
1			
2			
3			
4			
5			
6			
7			
8			
9			
10			

Activity 2: Now, take your groups from Activity 1 and *reduce them* to only three groups. That means you are going to have to think about how all 12 shapes can be put into *only* three different groups. The rule is the same as it was before. *You can use a shape only once.* That means that once you use a shape, you can't use it again.

Group	Label for Group	Numbers for the Shapes in That Group	Describe What Is the Same About the Shapes in This Group
1			
2			
3			

FINDING PRIME FACTORS FOR LARGE NUMBERS
More Geometry

LESSON 81

Name _____

PART A
WARM-UP ACTIVITY: BASIC AND EXTENDED FACTS

Activity: Solve the following basic and extended facts.

1. 300 ÷ 3 = _____
2. 6 × 7 = _____
3. 54 ÷ _____ = 6
4. 300 ÷ 10 = _____
5. 27 ÷ _____ = 9
6. _____ ÷ 7 = 5
7. 80 ÷ 2 = _____
8. 3 × _____ = 27
9. _____ ÷ 3 = 10

PART B
FINDING THE PRIMES OF LARGE NUMBERS

Activity: Use dividing rules and a calculator to find the prime factors for the following numbers.

_____ 108 _____

_____ 300 _____

_____ 135 _____

PART C
COMPARING SHAPES AGAIN

Activity: In the last lesson, you grouped similar shapes, with each shape appearing in *one group only*. This time, you are to group similar shapes again, only the shapes may appear in *more than one* group. In the table, make at least three groups for the shapes. Give a name to each shape group. Try to use each shape as many times as possible. For example, if you have grouped all of the shapes together that have straight edges, you might want to call that group "Straight Edges." Then write the numbers of the shapes in the column.

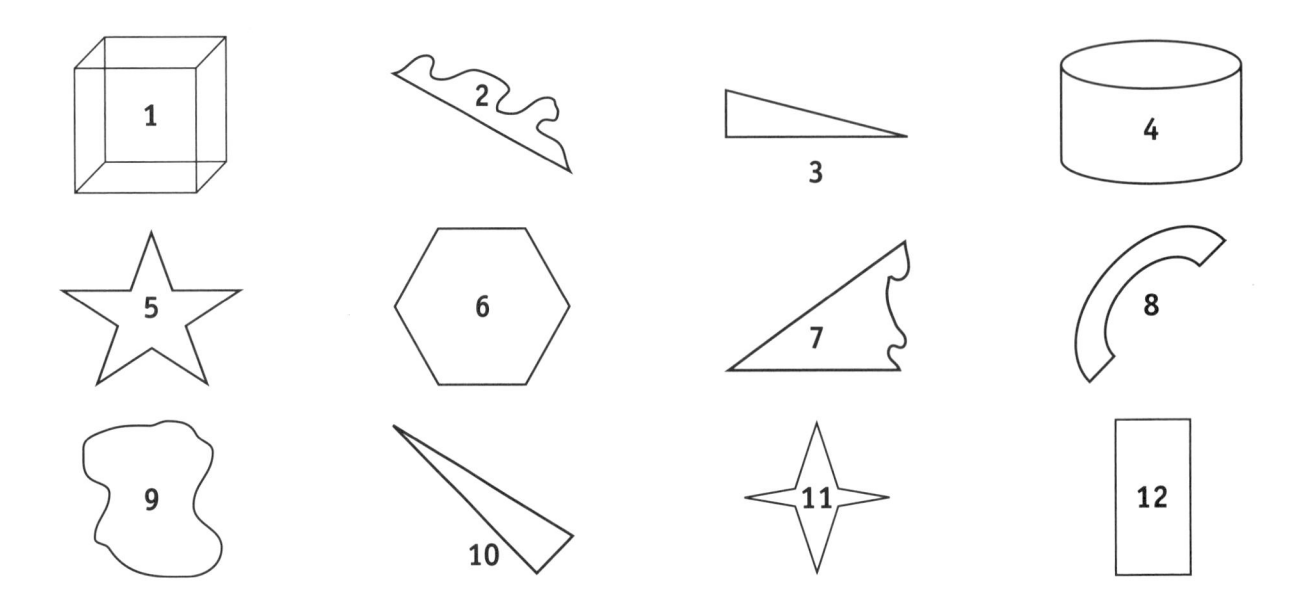

Group	Numbers for the Shapes in That Group	Describe What is the Same About the Shapes in This Group
1		
2		
3		
4		
5		
6		
7		
8		
9		
10		
11		
12		

FINDING COMMON FACTORS
More Patterns of Numbers

LESSON 82

Name _____

PART A
WARM-UP ACTIVITY: FACTORS

Activity: Circle all the factors for each of the following numbers.

1. 16 | 1 | 2 | 3 | 4 | 5 | 8 | 13 | 16 | 18 |

2. 21 | 1 | 2 | 3 | 5 | 7 | 9 | 11 | 21 | 24 |

3. 32 | 1 | 2 | 3 | 4 | 8 | 12 | 16 | 24 | 32 |

4. 49 | 1 | 3 | 5 | 7 | 9 | 11 | 22 | 33 | 49 |

5. 56 | 1 | 2 | 3 | 4 | 7 | 8 | 14 | 28 | 56 |

PART B
COMMON FACTORS

Activity: Do the following for the pairs of numbers in each problem.
1. Draw an "X" through the factors for each number.
2. Circle all of the factors the two numbers have in common.
3. List the numbers you have circled. These are the *common factors*.

1. 12 | 1 | 2 | 3 | 4 | 5 | 6 | 7 | 8 | 9 | 10 | 11 | 12 |
   15 | 1 | 2 | 3 | 4 | 5 | 6 | 7 | 8 | 9 | 10 | 11 | 12 | 13 | 14 | 15 |

The common factors are: _____

Student Workbook • *Number Sense*   Unit 5 Factors, Patterns, and Multiples   **229**

2.
8	1	2	3	4	5	6	7	8				
12	1	2	3	4	5	6	7	8	9	10	11	12

The common factors are: _____

3.
7	1	2	3	4	5	6	7						
14	1	2	3	4	5	6	7	8	9	10	11	12	13	14

The common factors are: _____

PART C
HOW MANY SQUARES?

Activity: This problem is based on a checkerboard or chessboard. How many squares do you see in the section that is circled? Be careful, the answer is not 16.

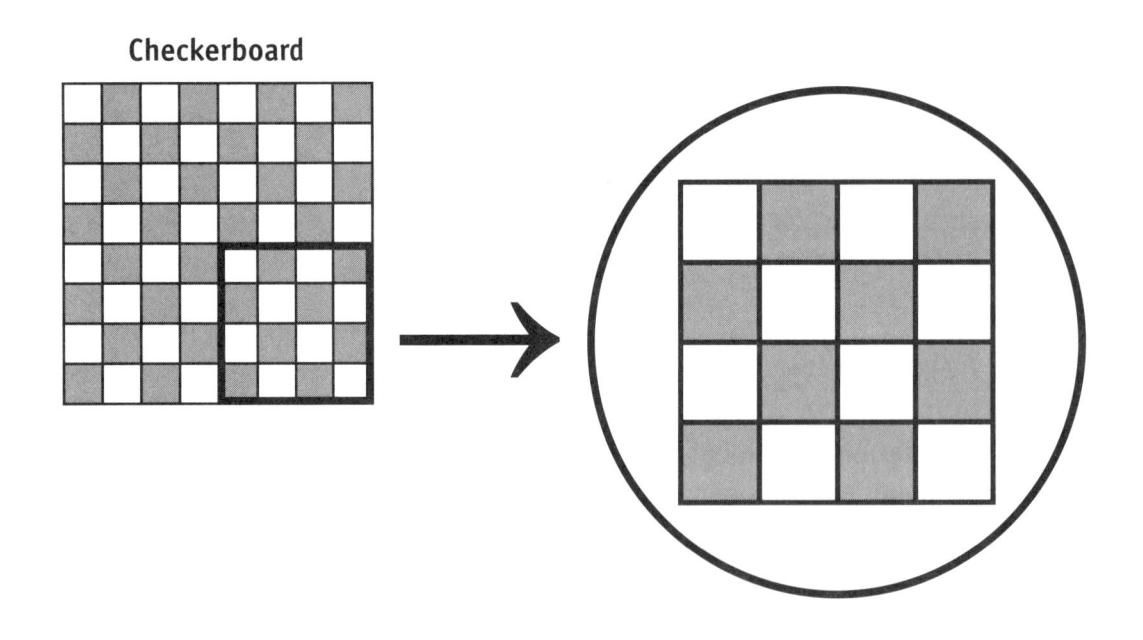

Checkerboard

How many squares are there? _____

How did you get your answer? _____

_____

_____

230   Unit 5 Factors, Patterns, and Multiples • Lesson 82   Student Workbook • *Number Sense*

MORE PRACTICE FINDING COMMON FACTORS
Geometry and Art

LESSON 83

Name _____

PART A
WARM-UP ACTIVITY: FACTORS

Activity: Circle all the factors for each of the following numbers.

1. 20 | 1 | 2 | 3 | 4 | 5 | 8 | 10 | 15 | 20 |

2. 18 | 1 | 2 | 3 | 4 | 6 | 9 | 12 | 15 | 18 |

3. 28 | 1 | 2 | 3 | 4 | 7 | 12 | 14 | 20 | 28 |

4. 42 | 1 | 2 | 3 | 4 | 6 | 7 | 14 | 21 | 42 |

5. 40 | 1 | 2 | 3 | 4 | 5 | 8 | 10 | 20 | 40 |

PART B
COMMON FACTORS

Activity: Do the following for the pairs of numbers in each problem.
1. Draw an "X" through the factors for each number.
2. Circle all of the factors the two numbers have in common.
3. List the numbers you have circled. These are the *common factors*.

1. 15 | 1 | 2 | 3 | 4 | 5 | 6 | 10 | 11 | 12 | 15 |
   20 | 1 | 2 | 3 | 4 | 5 | 6 | 10 | 11 | 12 | 15 | 16 | 17 | 18 | 19 | 20 |

The common factors are: _____

2.
18	1	2	3	4	5	6	8	9	10	11	14	16	18	
22	1	2	3	4	5	6	8	9	10	11	14	16	18	22

The common factors are: _____

3.
10	1	2	3	4	5	6	10							
30	1	2	3	4	5	6	10	12	15	20	22	25	27	30

The common factors are: _____

GREATEST COMMON FACTOR (GCF)
Looking at Patterns in Art

LESSON 84

Name _____

PART A
WARM-UP ACTIVITY: FINDING FACTORS

Activity: List all the factors for the following numbers.

1. The factors of 6 are: _____

2. The factors of 18 are: _____

3. The factors of 25 are: _____

4. The factors of 32 are: _____

PART B
GREATEST COMMON FACTOR

Activity: Do the following for the pairs of numbers in each problem.
1. List all of the factors for each number.
2. Circle all of the factors the two numbers have in common.
3. Tell which factor is the *greatest common factor*, or *GCF*.

1. 14
   21

What is the largest number circled? ____

We call this number the greatest common factor, or GCF.

2. 5
   20

What is the greatest common factor (GCF) of 5 and 20? ____

Student Workbook • *Number Sense*   Unit 5 Factors, Patterns, and Multiples   **233**

3. 14
   18

What is the greatest common factor (GCF) of 14 and 18? ____

4. 8
   12

What is the greatest common factor (GCF) of 8 and 12? ____

MORE PRACTICE WITH GCF
Congruent Shapes

LESSON 85

Name _____

PART A
WARM-UP ACTIVITY: FINDING FACTORS

Activity: List all the factors for the following numbers.

1. The factors of 9 are: _____
2. The factors of 12 are: _____
3. The factors of 16 are: _____
4. The factors of 25 are: _____

PART B
GREATEST COMMON FACTOR

Activity: Do the following for the pairs of numbers in each problem.
1. List all of the factors for each number.
2. Circle all of the factors the two numbers have in common.
3. Tell which factor is the *greatest common factor,* or *GCF.*

1. 10
   25

   What is the greatest common factor (GCF) of 10 and 25? ____

2. 6
   18

   What is the greatest common factor (GCF) of 6 and 18? ____

Student Workbook • *Number Sense*   Unit 5 Factors, Patterns, and Multiples   **235**

3.

16								
24								

What is the greatest common factor (GCF) of 16 and 24? ____

PART C
CONGRUENT SHAPES

Activity: Look at the shapes at the top of the grid. The shapes are a triangle, a parallelogram, and a rectangle. Each of these shapes was used to create *one* of the following designs labeled A, B, and C. The shape was copied, turned, and/or flipped to make the design. That shape, and *only that shape*, was used for the design. See if you can figure out which shape goes with which design. Draw the shapes in the designs.

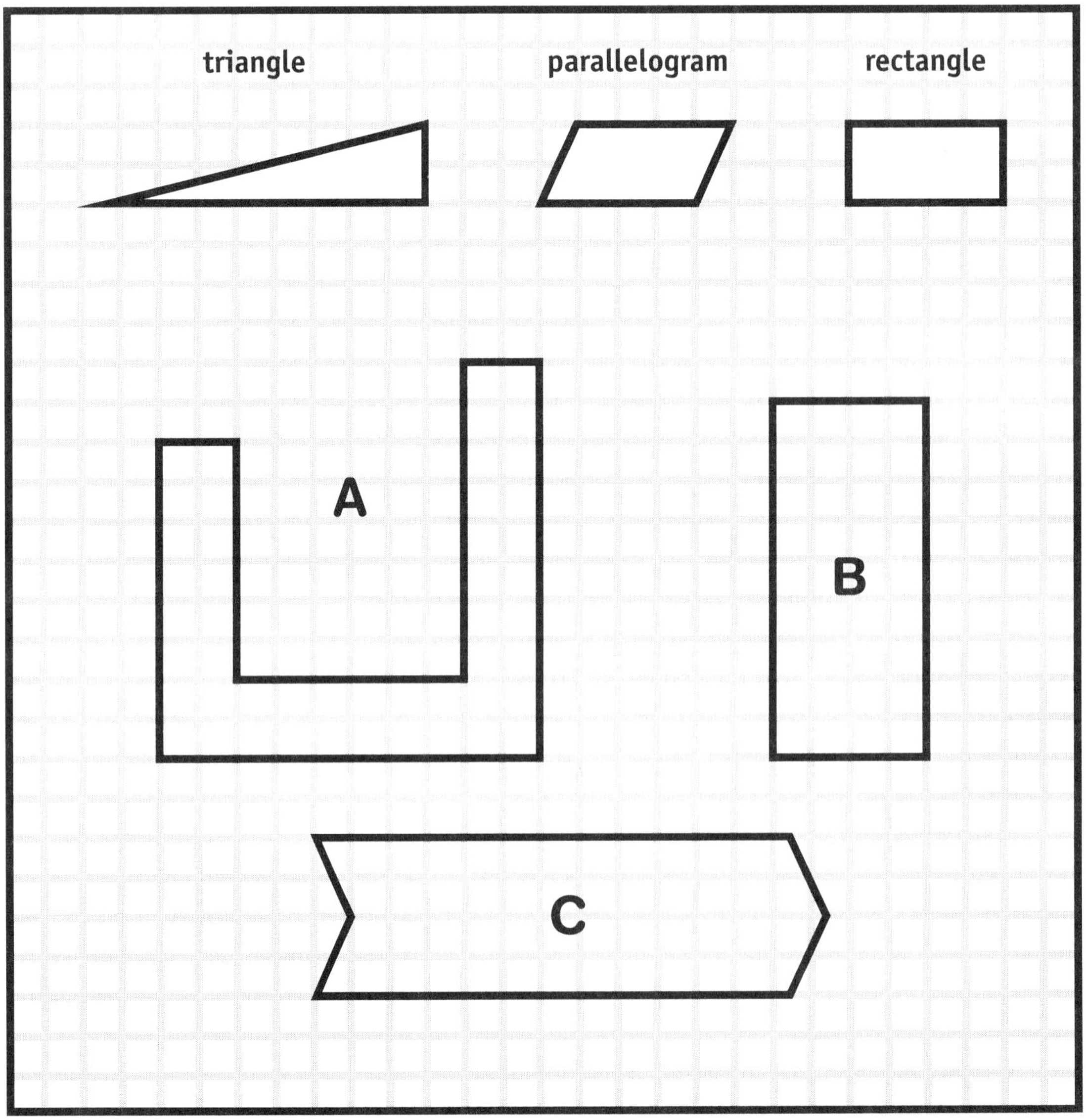

MAKING PATTERNS
Review of Prime Factor Trees

LESSON 86

Name _____

PART A
WARM-UP ACTIVITY: FINDING FACTORS

Activity: List all the factors for the following numbers.

1. The factors of 30 are: _____
2. The factors of 35 are: _____
3. The factors of 27 are: _____
4. The factors of 22 are: _____

PART B
REVIEW: PRIME FACTOR TREES

Activity: Make prime factor trees for these two large numbers. Use dividing rules and a calculator to find the prime factors.

_____ 66 _____   _____ 90 _____

Student Workbook • *Number Sense*   Unit 5 Factors, Patterns, and Multiples   **237**

PART C
BUILDING PATTERNS

Activity: In this activity you are going to use a triangle to explore patterns. How many *different shapes* can you make by putting together five triangles in a variety of ways? Here's the rule: One side of each triangle must lie flat against a side of another triangle. We have provided two grid pages for you to use.

← sides must touch

We have drawn the first triangle for you.

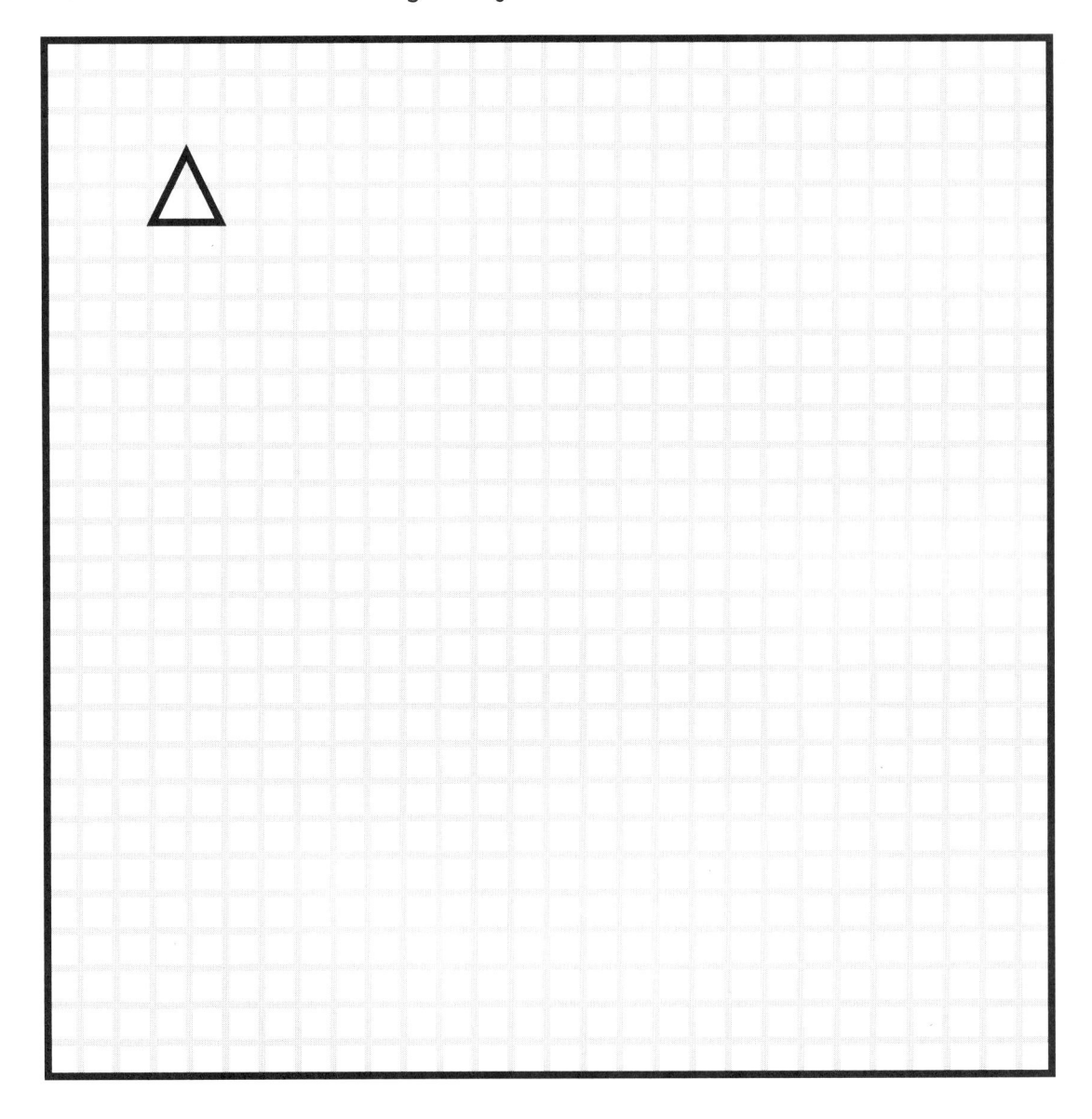

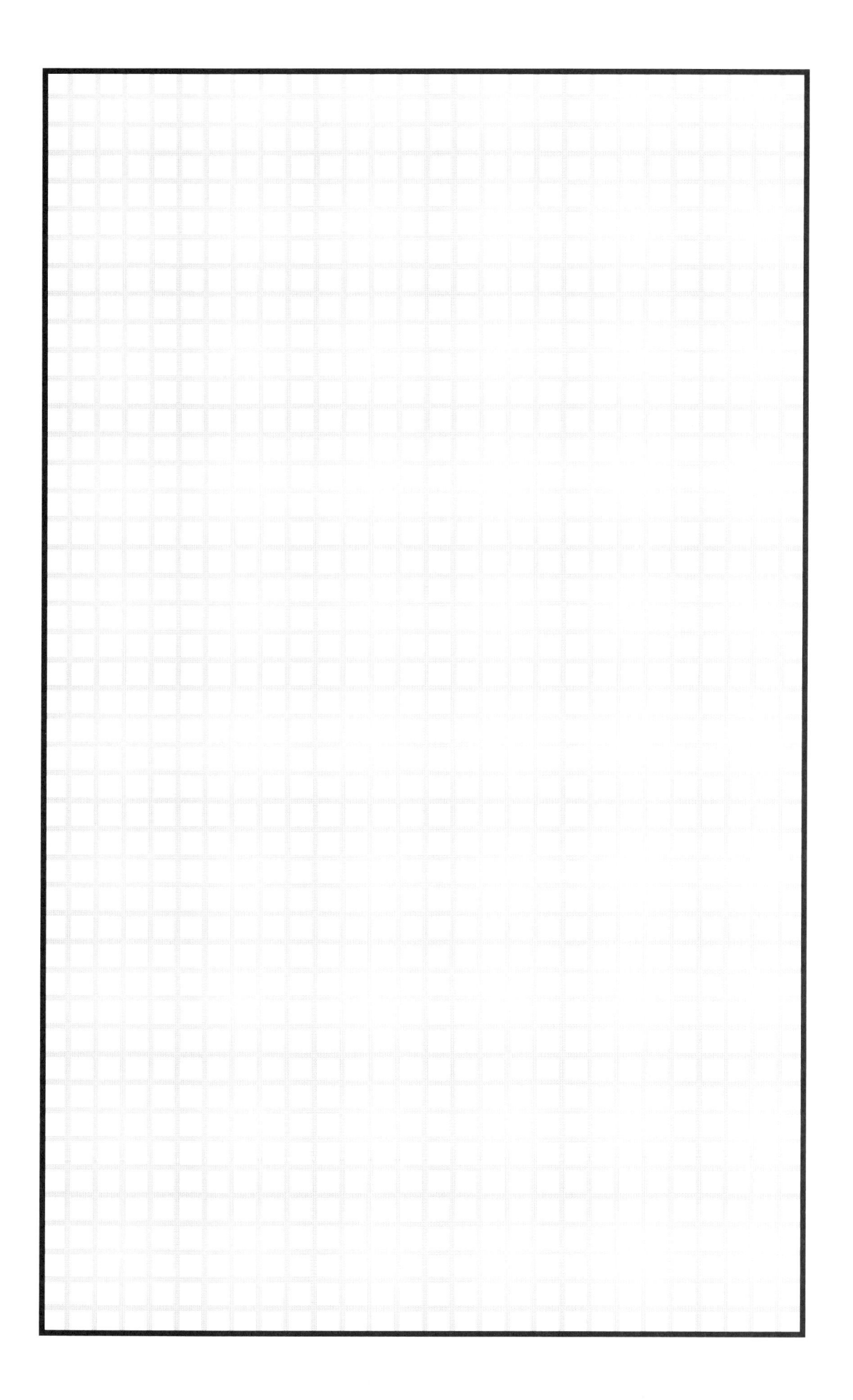

CHALLENGE LESSON: STRATEGIES FOR FINDING GCF
Using Prime Factor Trees

LESSON 87

Name _____

PART A
WARM-UP ACTIVITY: DIVISION FACTS

Activity: Solve the following basic facts.

1. $30 \div 3 =$ ____
2. $9 \times 9 =$ ____
3. $42 \div$ ____ $= 6$
4. $35 \div 7 =$ ____
5. $24 \div$ ____ $= 6$
6. ____ $\div 8 = 8$
7. $50 \div 10 =$ ____
8. $4 \times$ ____ $= 36$
9. ____ $\div 7 = 7$

PART B
GCF FOR LARGE NUMBERS

Activity: Find the greatest common factor of these two numbers. Use prime factor trees to find the GCF.

1. Find the greatest common factor of 65 and 75.

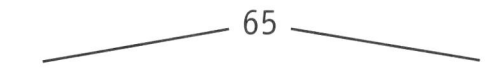

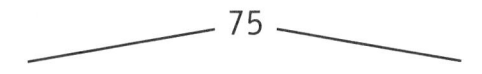

_____ 65 _____   _____ 75 _____

Circle all the primes. Then put a check mark by the ones that match *on both sides*.

Multiply all of the numbers that you have checked to get the greatest common factor. If there's only one number, then that is the greatest common factor.

The greatest common factor of 65 and 75 is: ____

Student Workbook • Number Sense   Unit 5 Factors, Patterns, and Multiples   241

2. Find the greatest common factor of 90 and 105.

_____ 90 _____   _____ 105 _____

Circle all the primes. Then put a check mark by the ones that match *on both sides*.

Multiply all of the numbers that you have checked to get the greatest common factor. If there's only one number, then that is the greatest common factor.

The greatest common factor of 90 and 105 is: _____

PART C
CONGRUENT TRIANGLES

Activity: Break the following shapes into congruent triangles. The triangles inside each shape must be the same size.

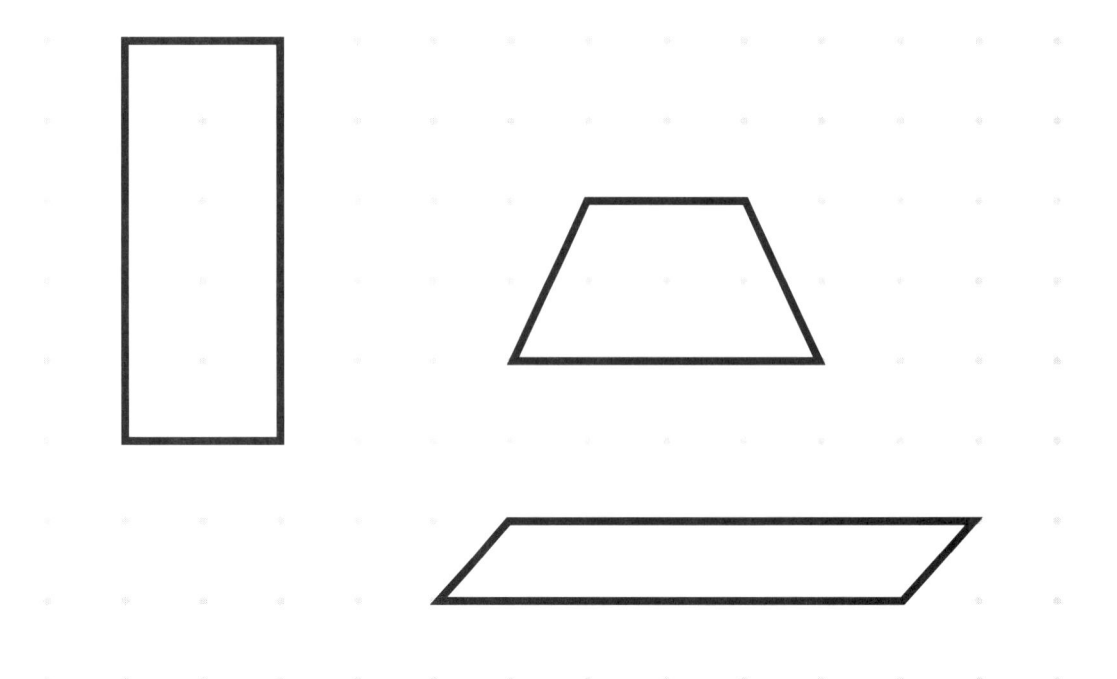

REVIEW OF LESSONS 79 TO 87

Name _____

LESSON 88

PART A
DIVIDING RULES

Here is a list of the dividing rules you have studied in this unit:

2	We can divide a number by 2 if it is an even number.
3	We can divide a number by 3 if we can add up its digits and divide that number by 3 evenly.
5	We can divide a number by 5 if it ends in a 5 or a 0.
6	We can divide a number by 6 if we can divide it by 2 *and* by 3.
10	We can divide a number by 10 if it ends in a 0.

Activity: Which *rule* or *rules* work for the following numbers? The rules for the first number have been filled in for you.

Number	Rule or Rules
270	2, 3, 5, 6, and 10 rules
475	
76	
873	
42,010	
822	

Student Workbook • *Number Sense*   Unit 5 Factors, Patterns, and Multiples   243

PART B
USING AREA FORMULAS

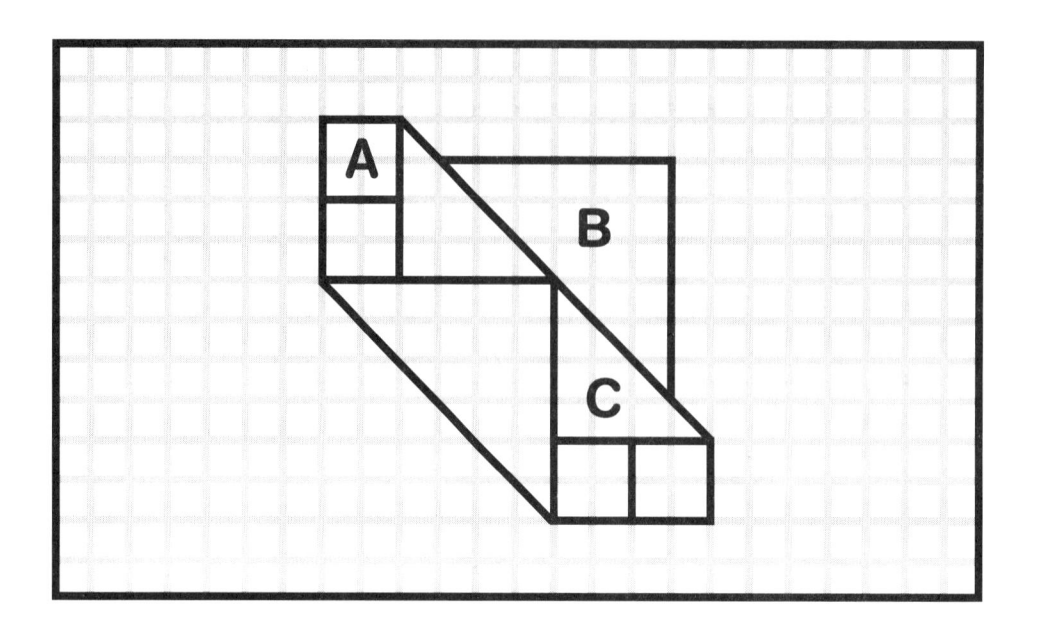

Activity: Find the following areas. Remember that area is measured in square units.

1. What is the area of object A? ____

2. What is the area of object B? ____

3. What is the area of object C? ____

PART C
PRIMES FOR LARGE NUMBERS

Activity: Use dividing rules and a calculator to find the prime factors of the following large numbers.

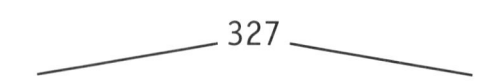

1. 327

2. ⌒ 426 ⌒

3. ⌒ 468 ⌒

PART D
COMMON FACTORS

Activity: Do the following for the pairs of numbers in each problem.
1. Draw an "X" through the factors for each number.
2. Circle all of the factors the two numbers have in common.
3. List the numbers you have circled. These are the *common factors*.

1.   5

1	2	3	4	5										

   20

1	2	3	4	5	6	10	11	12	15	16	17	18	19	20

The common factors are: _____

2.   7

1	2	3	4	5	6	7							

   21

1	2	3	4	5	6	7	11	12	15	16	17	19	20	21

The common factors are: _____

3.   8

1	2	3	4	5	6	7	8						

   24

1	2	3	4	5	6	7	8	12	16	18	20	22	23	24

The common factors are: _____

PART E
GREATEST COMMON FACTOR

Activity: Do the following for the pairs of numbers in each position.
1. List all of the factors for each number.
2. Circle all of the factors the two numbers have in common.
3. Tell which factor is the *greatest common factor*, or *GCF*.

1. 12
   24

   What is the greatest common factor (GCF) of 12 and 24? _____

2. 16
   32

   What is the greatest common factor (GCF) of 16 and 32? _____

3. 20
   40

   What is the greatest common factor (GCF) of 20 and 40? _____

PART F
CONGRUENT SHAPES

Activity: Tell which of these shapes are congruent.

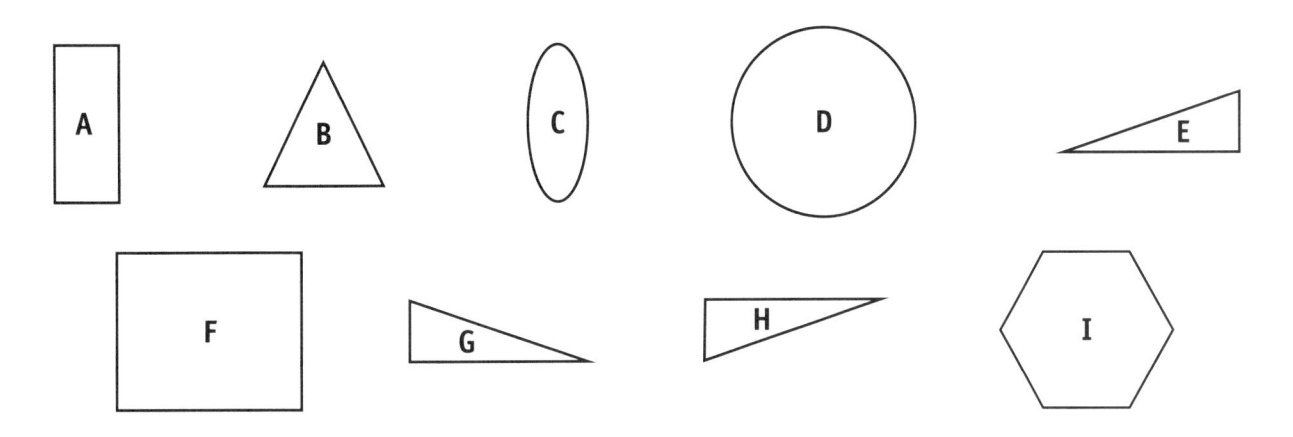

PART G
GCF FOR LARGE NUMBERS

Activity: Find the greatest common factor for these two numbers. Use prime factor trees to find the GCF.

Find the greatest common factor of 35 and 75.

Circle all the primes. Then put a check mark by the ones that match *on both sides*. Multiply all of the numbers that you have checked to get the greatest common factor. If there's only one number checked, then that is the greatest common factor.

The greatest common factor of 35 and 75 is: _____

PATTERNS OF NUMBERS: EVENS AND ODDS
Expanding and Contracting Objects

LESSON 89

Name _____

PART A
WARM-UP ACTIVITY: EXTENDED FACTS

Activity: Solve the following extended facts.

1. 30 + 80 = ____
2. 900 + 600 = ____
3. 30 × 7 = ____
4. 80 + ____ = 140
5. 50 × 10 = ____
6. 4 × ____ = 400

PART B
EVEN AND ODD NUMBERS

Activity: Think about the kinds of numbers you will get when you add or multiply the numbers in the chart. *Do not work the problems.* Just put an X in the box to indicate whether the answer to the problem will be even or odd. The first problem is done for you.

Problem	The Answer Will Be: Even Number	Odd Number
6,349 + 76,945	X	
987 + 1,854		
889 × 935		
93,456 + 123,554		
773 + 595		
87 × 359		
328,340 + 222		
132,602 × 67		
22,421 + 648,448		
393 × 107		

Student Workbook • *Number Sense* Unit 5 Factors, Patterns, and Multiples **249**

PART C
EXPANDING AND CONTRACTING COMMON OBJECTS

Activity 1: Each square in this grid has a base of 1 cm and a height of 1 cm. Draw a rectangle that has a base of 4 cm and a height of 2 cm. Next, make two more rectangles, as follows:

One rectangle should have a base and height that are both 3 times larger than in your original rectangle. The next rectangle should have a base and height that are both half as large as in your original rectangle.

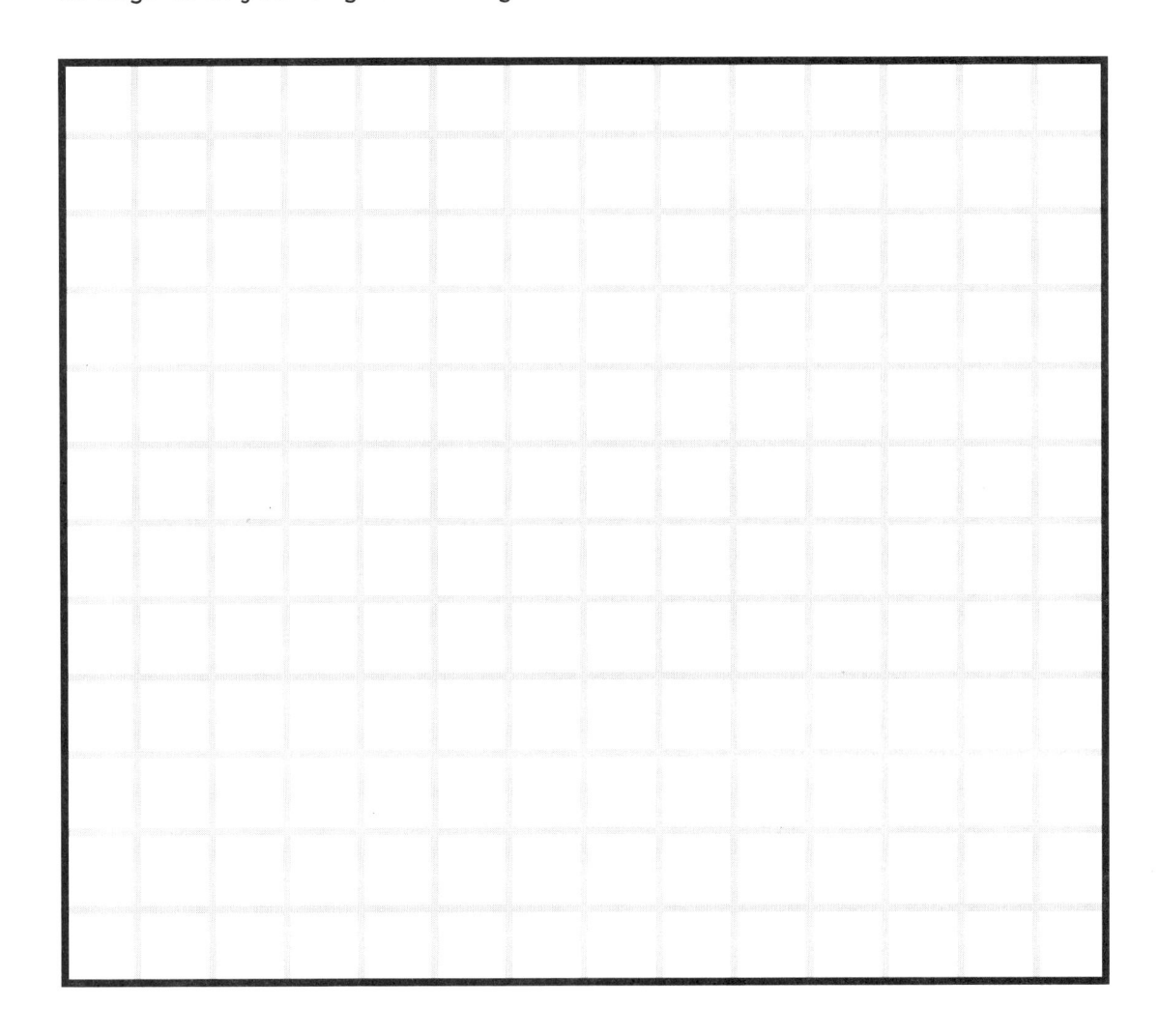

Activity 2: Label your rectangles 1, 2, and 3. Then fill in the following table. Make sure to give units for perimeter and area. Remember, the area should be in square cm (sq. cm). After you fill in the table, answer the questions that follow it.

Rectangle	Perimeter	Area
1		
2		
3		

1. What happens to the perimeter when you go from Rectangle 1 to Rectangle 2?

2. What happens to the area when you go from Rectangle 1 to Rectangle 2?

3. How much bigger is the area of Rectangle 2 than the area of Rectangle 3?

PATTERNS OF NUMBERS
Square Numbers

LESSON 90

Name _____

PART A
WARM-UP ACTIVITY: BASIC AND EXTENDED FACTS

Activity: Solve the following basic and extended facts.

1. $9 \times 9 = $ ____
2. $80 \times 8 = $ ____
3. $300 \times 3 = $ ____
4. $7 \times 7 = $ ____
5. $60 \times 6 = $ ____
6. $900 \times 9 = $ ____
7. $5 \times 5 = $ ____
8. $40 \times 4 = $ ____
9. $700 \times 7 = $ ____

PART B
EXPANDING SQUARE NUMBERS

An interesting way to investigate square numbers is through geometry. In Lesson 89, you expanded and contracted rectangles. In this activity, you will look at what happens to the area of a square when you increase the length of the base and height of the square by 1 cm each time. You will be making a new kind of design, one that looks like this:

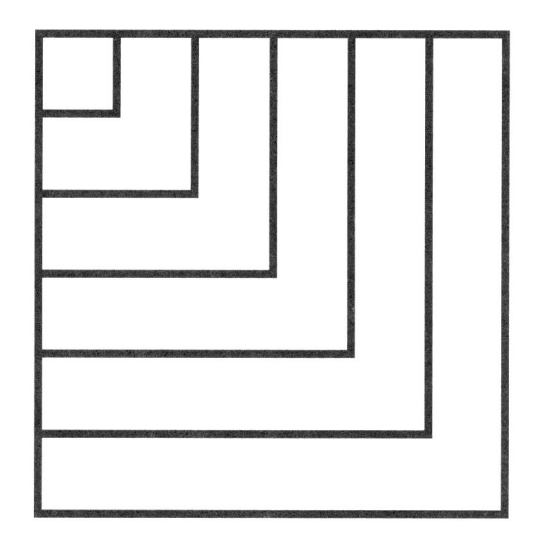

Student Workbook • *Number Sense* Unit 5 Factors, Patterns, and Multiples   253

Activity 1: Start with the square in the upper left corner of the grid. It has a base of 1 cm and a height of 1 cm.

Now draw a square in the upper left corner of the grid that has a base that is 1 cm bigger and a height that is 1 cm bigger than the first square. The new square will have a base of 2 cm and a height of 2 cm. The squares should overlap, as shown in the design on the previous page.

Keep going until you have drawn 10 squares, each one 1 cm in base and 1 cm in height bigger than the one before it.

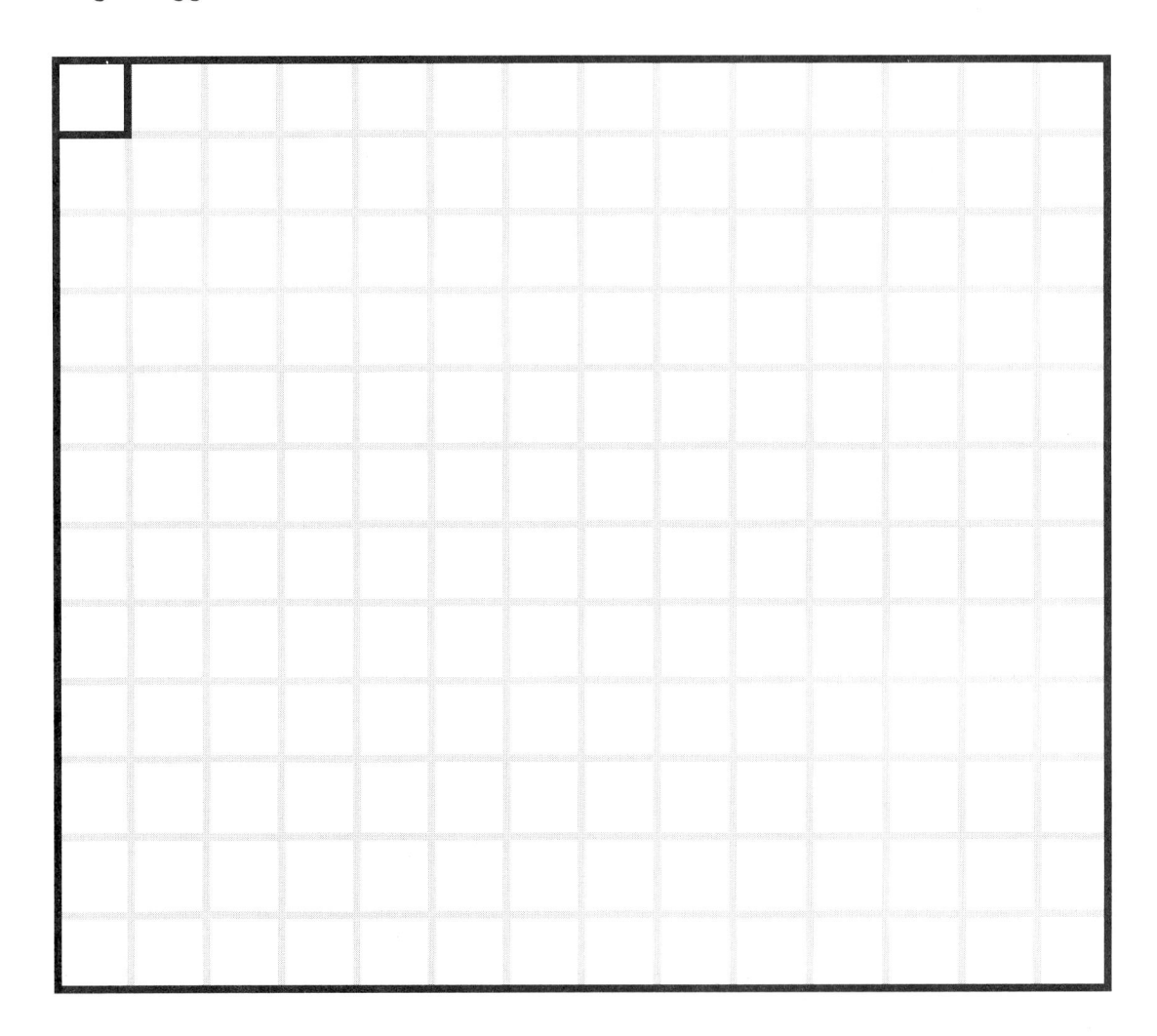

Activity 2: List in this table the size of each square from Activity 1. Fill in the area. Then find the difference in area between each two consecutive squares. When you are done, answer the questions that follow the table.

Size of Square	Area	Difference Between Two Squares
1 × 1	1 sq. cm	
		3 sq. cm
2 × 2	4 sq. cm	

Answer the following questions about the activity above:

1. What patterns do you see in the table and organized list that you have created?

_____

_____

2. What would be the difference in area between the next two consecutive squares (that is, between 10 × 10 and 11 × 11)?

_____

PATTERN OF NUMBERS: SQUARE NUMBERS AND ODD NUMBERS

Expanding Triangles

LESSON 91

Name _____

PART A
WARM-UP ACTIVITY: FACTOR RAINBOWS

Activity: Draw factor rainbows for the following numbers and then list all of the factors.

1. Draw the factor rainbow for the number 25.

    The factors for 25 are: _____

2. Draw the factor rainbow for the number 32.

    The factors for 32 are: _____

3. Draw the factor rainbow for the number 23.

    The factors for 23 are: _____

PART B
USING BLOCKS TO PICTURE SQUARE NUMBERS

Activity: Continue adding square blocks to the picture below so that you get bigger and bigger square arrays. Each time you make a square array, go to the table that follows and fill in the following information:

- the number the array represents
- description of the array
- the number of blocks in the array
- the number of blocks you added to make the square number

The beginning of the table has been filled in for you.

Number the Array Represents	Description of the Array	Number of Blocks in the Array	Number of Blocks You Added
1	1 × 1	1	
4	2 × 2	4	3

PART C
EXPANDING TRIANGLES

Activity: You have already expanded squares and rectangles. Now it is time to expand a triangle. Look at the grid on this page. Each square on the grid has a base of 1 cm and a height of 1 cm. Now look closely at the triangle. In another space on the grid, make another triangle that has a base that is 3 times bigger and a height that is 2 times bigger. Then answer the questions that follow the grid.

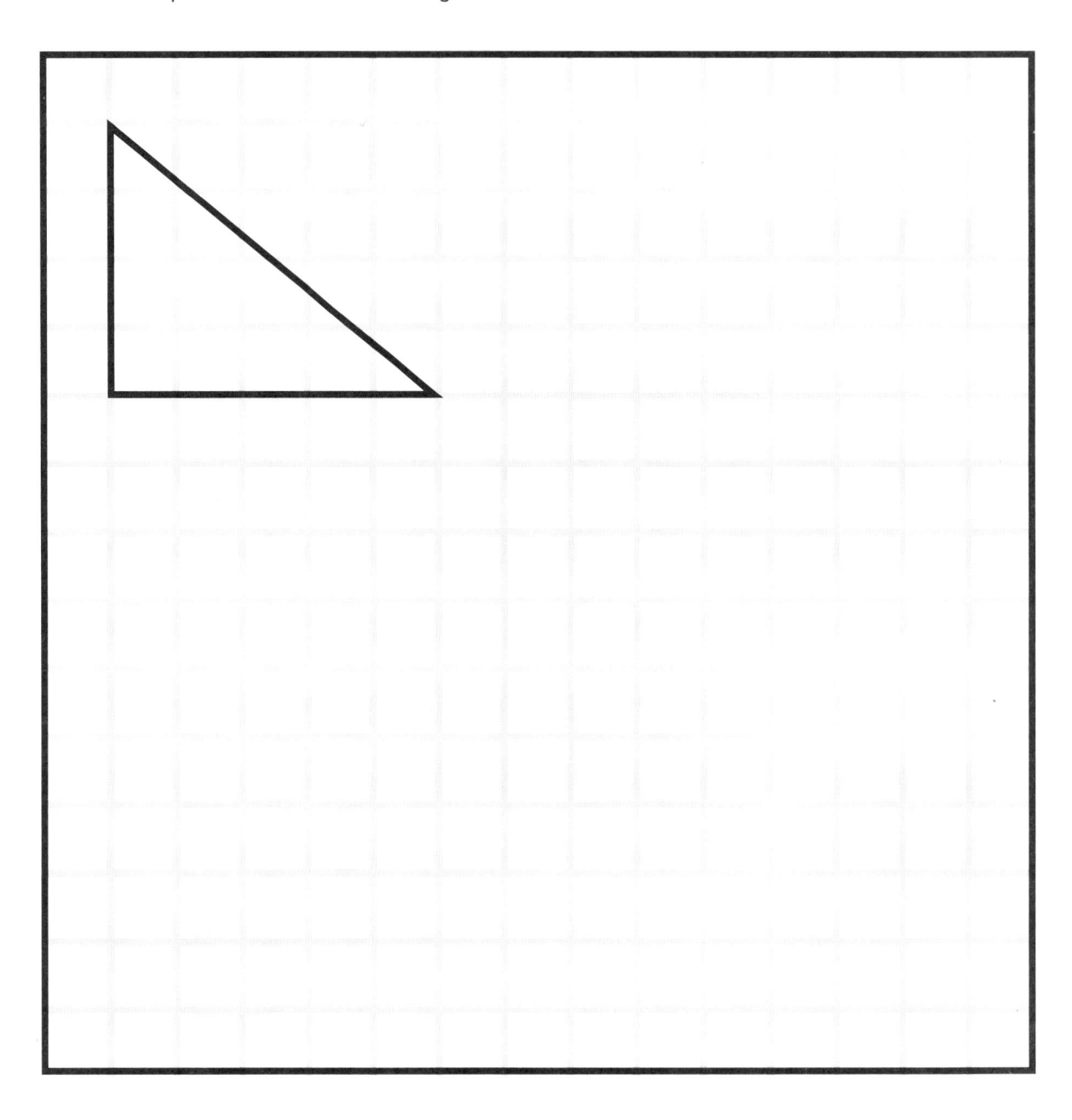

What is the *area* of the original triangle? _____

What is the *area* of the triangle that you drew? _____

PATTERNS OF NUMBERS: TRIANGULAR NUMBERS

Problem Solving: Finding Patterns

Name _____

LESSON 92

PART A
WARM-UP ACTIVITY: ADDITION

Activity: Solve the following addition problems. The answers are related to a pattern you will learn in today's lesson—triangular numbers.

1. 1 + 2 = _____
2. 10 + 5 = _____
3. 28 + 8 = _____
4. 3 + 3 = _____
5. 15 + 6 = _____
6. 36 + 9 = _____
7. 6 + 4 = _____
8. 21 + 7 = _____
9. 45 + 10 = _____

PART B
WORKING WITH TRIANGULAR NUMBERS

Activity: Copy your answers to Problems 1 through 9 in Part A in the left column of this table. Use your calculator (multiply × 8 and add 1) to find the square number. Then use your calculator to multiply the numbers in the last column. Do the two square numbers match? The first row has been filled in for you.

Answers from Part A	8 Times Your Answer Plus 1	Check to See if the Number in Column 2 Is a Square
1. 3	3 × 8 = 24 + 1 = **25**	5 × 5 = **25**
2.		11 × 11 =
3.		17 × 17 =
4.		7 × 7 =
5.		13 × 13 =
6.		19 × 19 =
7.		9 × 9 =
8.		15 × 15 =
9.		21 × 21 =

Student Workbook • Number Sense

PATTERNS OF NUMBERS: WHEN NUMBERS REPEAT
Scaling: Expanding and Contracting

LESSON 93

Name _____

PART A
WARM-UP ACTIVITY: PRIME FACTORIZATION

Activity 1: Fill in the missing values in these prime factor trees. Then fill in the prime factorizations.

1.

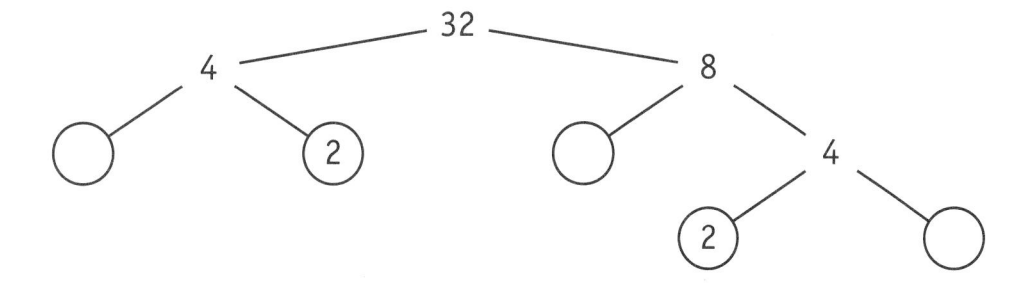

The prime factorization of 32 is: ____ × ____ × ____ × ____ × ____.

2.

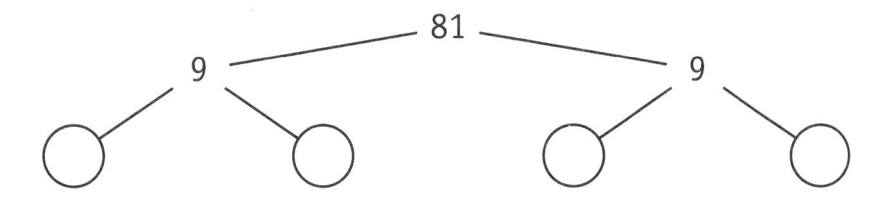

The prime factorization of 81 is: ____ × ____ × ____ × ____.

Student Workbook • *Number Sense*   Unit 5 Factors, Patterns, and Multiples   263

PART B
EXPONENTS

Activity 1: Rewrite the following multiplication problems using exponents.

EXAMPLE $2 \times 2 \times 2 \times 3 \times 3 \times 3 \times 3 = $ ___

Answer: $2^3 \times 3^4$

1. $5 \times 5 \times 5 \times 5 = $ _____
2. $2 \times 2 \times 2 \times 3 \times 5 = $ _____
3. $2 \times 2 \times 3 \times 3 \times 5 \times 7 = $ _____
4. $10 \times 10 \times 10 \times 10 \times 10 \times 10 \times 10 \times 10 \times 10 = $ _____
5. $3 \times 3 \times 7 = $ _____

Activity 2: Expand the following numbers.

EXAMPLE $2^5 = $ _____

Answer: $2 \times 2 \times 2 \times 2 \times 2$

1. $3^2 = $ _____
2. $7^4 = $ _____
3. $9^3 = $ _____
4. $2^2 \times 3^2 = $ _____
5. $5^3 \times 4^2 = $ _____

PART C
SCALING OBJECTS TO FIT A BIGGER AREA

Activity: A company wants to have its Web page look like the design it uses on its business cards. We have copied the business card onto a grid with squares that you should assume have a base of 1 cm and a height of 1 cm.

Use the space below the business card to expand this design. All of the objects in the design should be 2 times as big as they are in the drawing of the business card. Hint: Start your drawing in the middle of the grid so you have enough room. You do not need to include the company name.

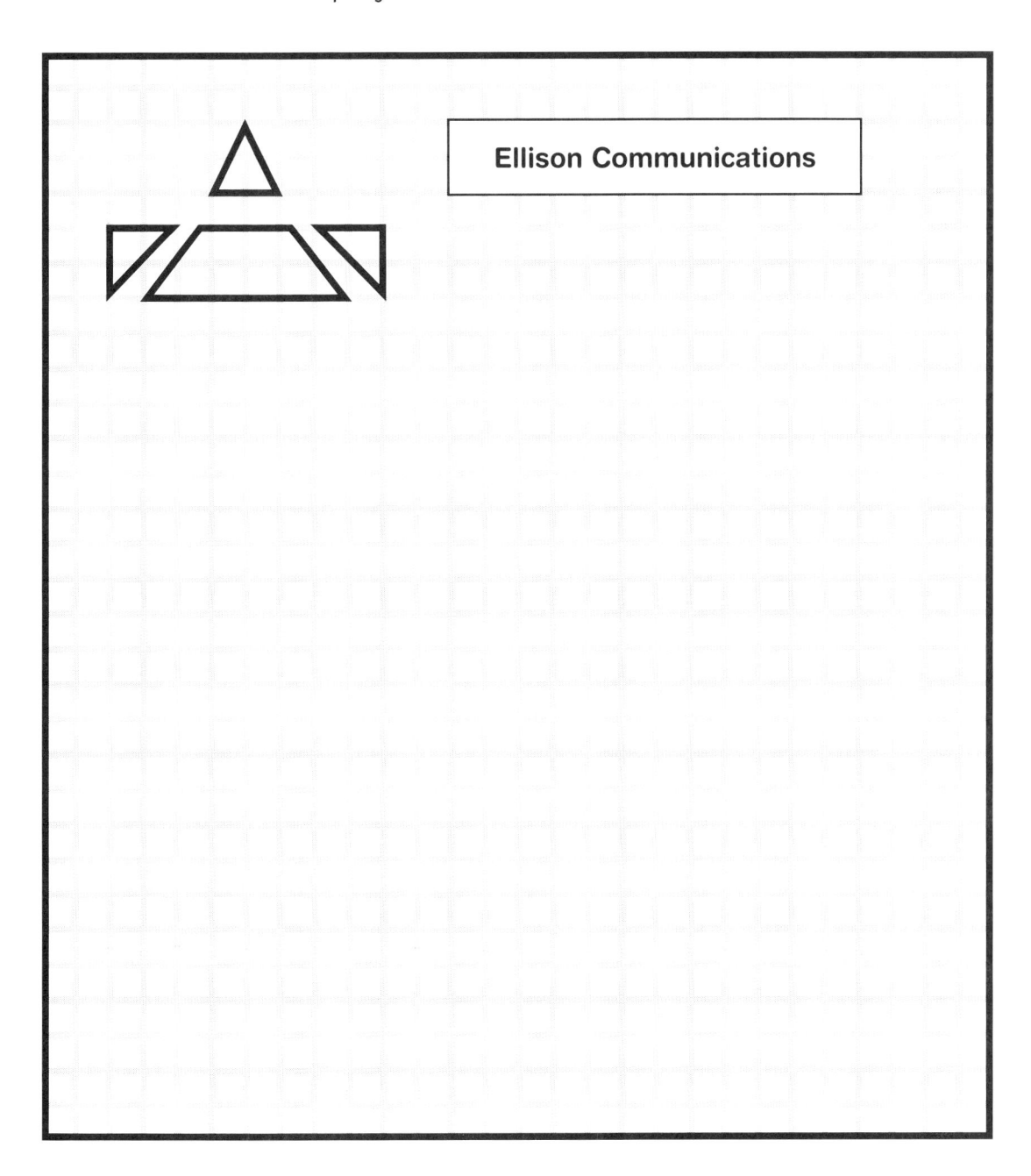

Ellison Communications

Student Workbook • *Number Sense*   Unit 5 Factors, Patterns, and Multiples • Lesson 93

COMMON MULTIPLES
Changing Objects

LESSON 94

Name _____

PART A
WARM-UP ACTIVITY: EXPONENTS

Activity: Rewrite the following numbers using exponents.

$$2 \times 2 \times 2 \times 2 \times 2 = 2^5$$ **EXAMPLE**

1. $4 \times 4 \times 4 \times 4 \times 4 =$ _____

2. $30 \times 30 \times 30 \times 30 \times 30 =$ _____

3. $8 \times 8 \times 3 \times 3 =$ _____

4. $10 \times 10 \times 10 \times 9 \times 9 \times 9 \times 9 \times 9 \times 9 =$ _____

5. $6 \times 8 \times 8 =$ _____

PART B
MULTIPLES

Activity 1: Write the first five *multiples* of each number. Start each list by rewriting the number itself.

         6:   6   12   18   24   30   **EXAMPLE**

1. 7: ____ ____ ____ ____ ____

2. 9: ____ ____ ____ ____ ____

3. 5: ____ ____ ____ ____ ____

4. 2: ____ ____ ____ ____ ____

5. 10: ____ ____ ____ ____ ____

Student Workbook • *Number Sense*   Unit 5 Factors, Patterns, and Multiples   **267**

Activity 2: We have drawn pairs of number lines for each problem. Using the number lines, find common multiples for each pair of numbers and list them.

1. Find common multiples of 5 and 7. They include: _____

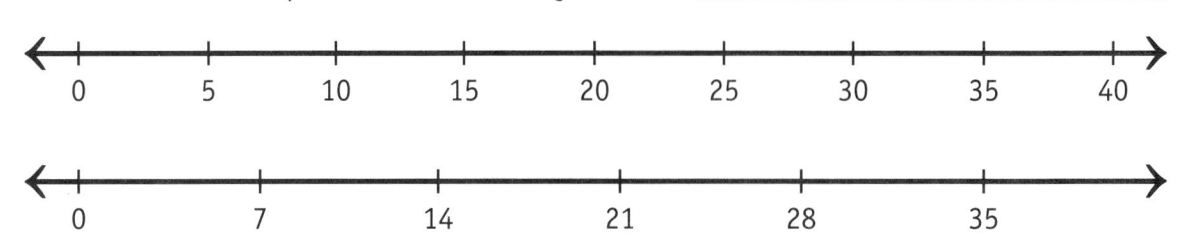

2. Find common multiples of 2 and 3. They include: _____

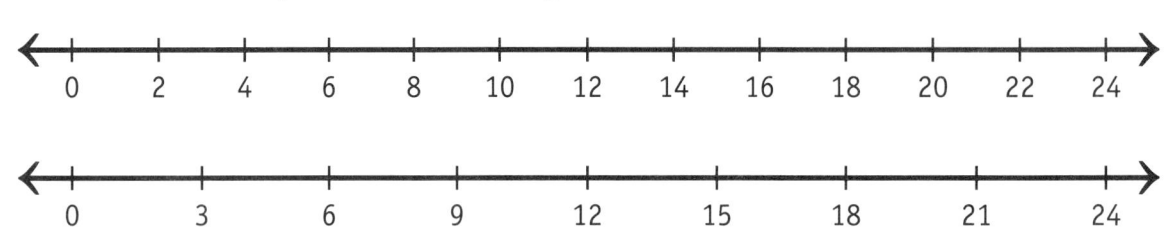

3. Find common multiples of 5 and 10. They include: _____

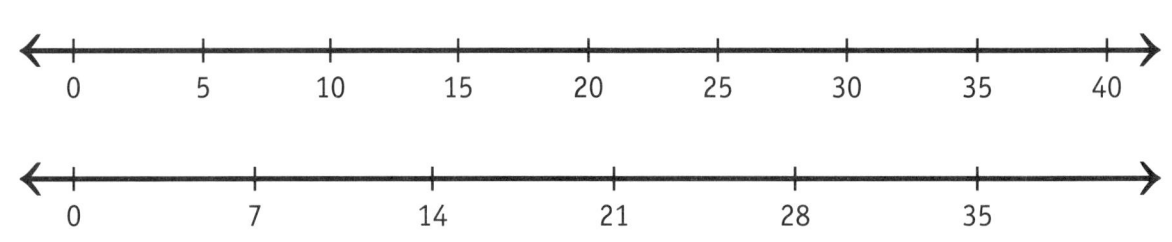

4. Find common multiples of 6 and 8. They include: _____

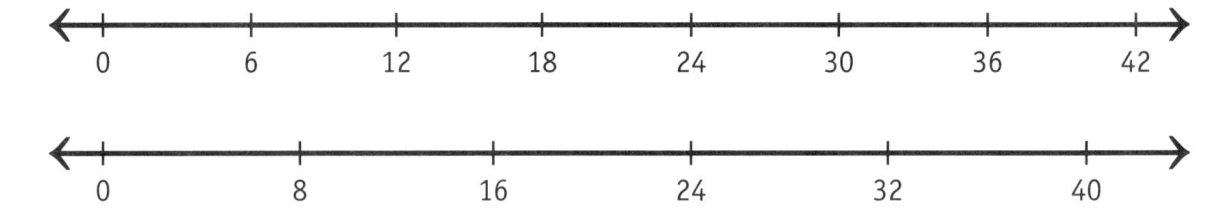

5. Find common multiples of 7 and 3. They include: _____

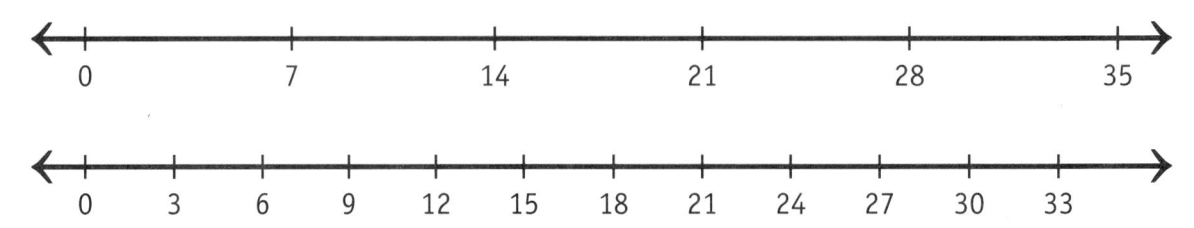

PART C
SLIDE, FLIP, OR TURN?

Activity 1: We have moved all of the following objects. Tell whether each move was a slide, flip, or turn.

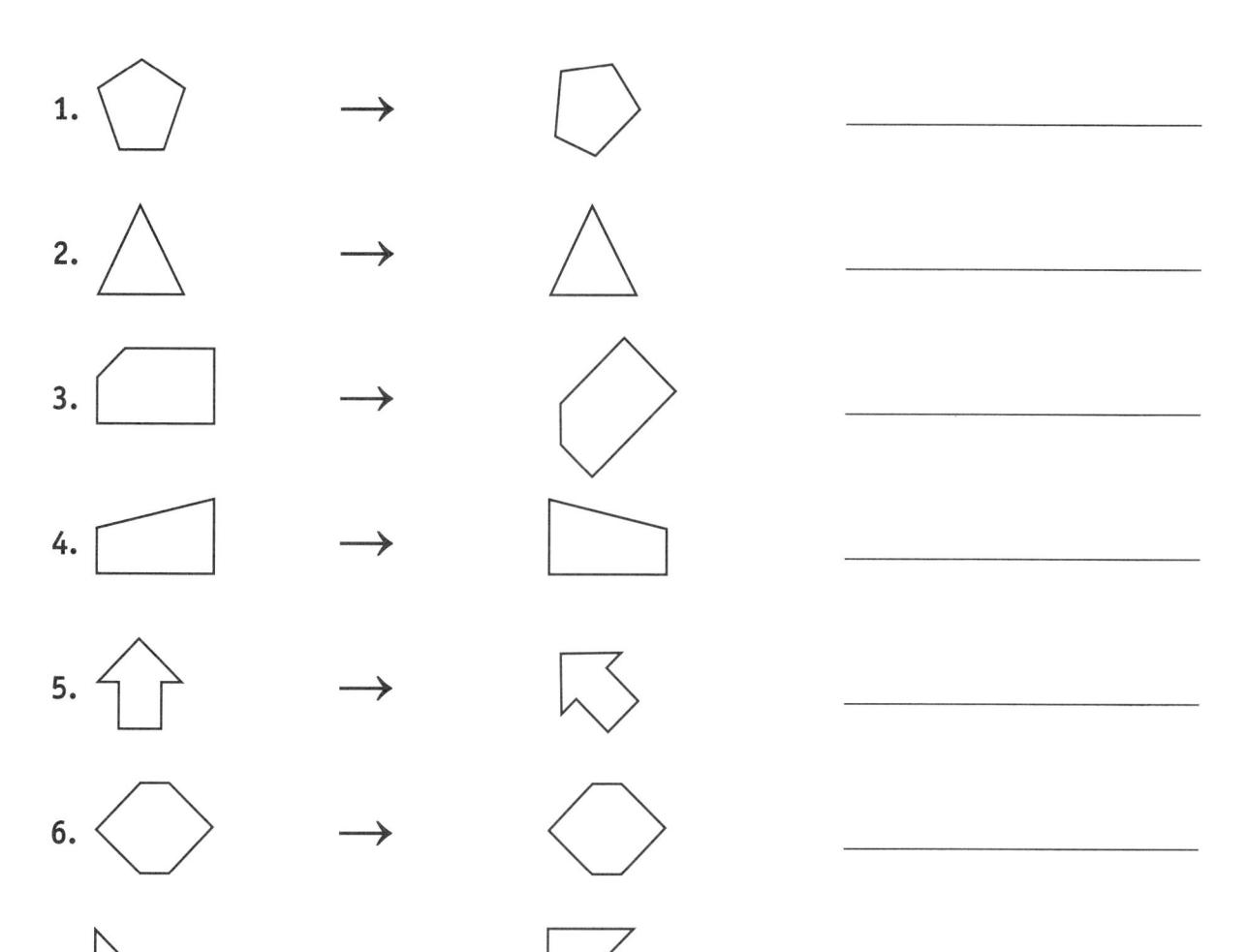

1. _____

2. _____

3. _____

4. _____

5. _____

6. _____

7. _____

Activity 2: Draw two different objects on the grid. Then show what each object would look like when you:
1. Slide it
2. Flip it
3. Turn it

FINDING THE LEAST COMMON MULTIPLE (LCM)
Symmetry

LESSON 95

Name _____

PART A
WARM-UP ACTIVITY: EXPONENTS AND MULTIPLES

Activity 1: Rewrite the following numbers using exponents.

$4 \times 4 \times 4 =$ _____   **EXAMPLE**

Answer: 4^3

1. $5 \times 5 \times 3 \times 3 \times 3 =$ _____
2. $6 \times 6 \times 6 \times 6 \times 4 =$ _____
3. $7 \times 9 \times 9 \times 9 \times 9 =$ _____
4. $8 \times 8 \times 8 \times 7 \times 7 \times 7 \times 6 \times 6 \times 6 =$ _____
5. $3 \times 3 \times 100 \times 100 =$ _____

Activity 2: Write the first five *multiples* of each number. Start each list by rewriting the number itself.

7:   7   14   21   28   35   **EXAMPLE**

1. 5: ____ ____ ____ ____ ____
2. 3: ____ ____ ____ ____ ____
3. 8: ____ ____ ____ ____ ____
4. 9: ____ ____ ____ ____ ____
5. 4: ____ ____ ____ ____ ____

PART B
LEAST COMMON MULTIPLE OF TWO NUMBERS

Activity: Make organized lists to find the least common multiple (LCM) of the following numbers. When you find the least common multiple, circle it.

1. What is the *least common multiple (LCM)* of 5 and 10? _____

   5
   10

2. What is the *least common multiple (LCM)* of 2 and 7? _____

   2
   7

3. What is the *least common multiple (LCM)* of 4 and 6? _____

   4
   6

4. What is the *least common multiple (LCM)* of 6 and 8? _____

   6
   8

5. What is the *least common multiple (LCM)* of 3 and 4? _____

   3
   4

PART C
SYMMETRY IN OBJECTS

Activity: Answer the two questions that follow the grid. If a line drawn *does not* show symmetry, redraw it so that it does.

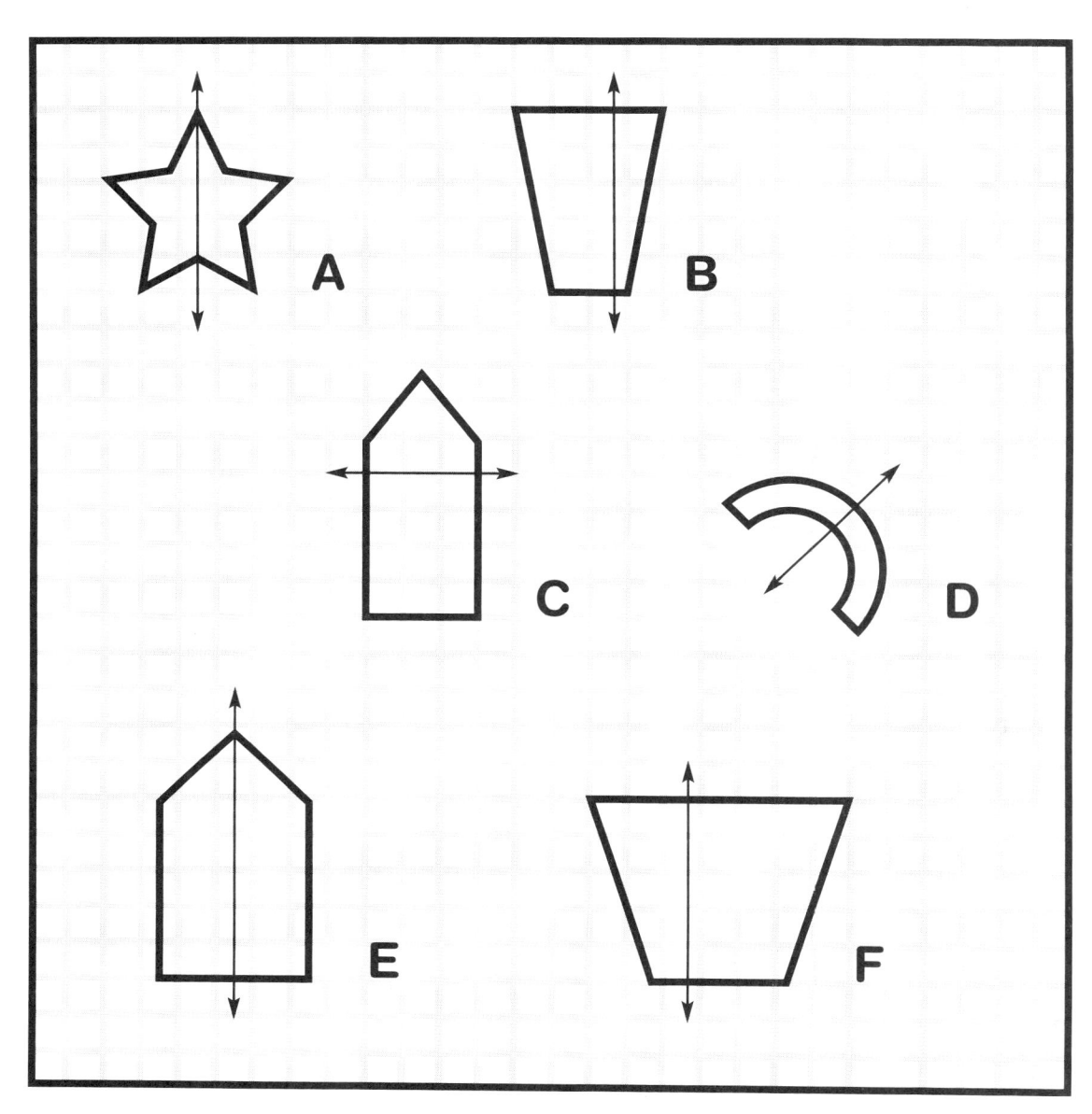

Which objects have lines that show symmetry? _____

Which objects have lines that *do not* show symmetry? _____

MORE PRACTICE FINDING LCM
Wheels Inside a Cuckoo Clock

LESSON 96

Name _____

PART A
WARM-UP ACTIVITY: PRIME FACTOR TREES

Activity 1: Fill in the missing values in the following prime factor trees.

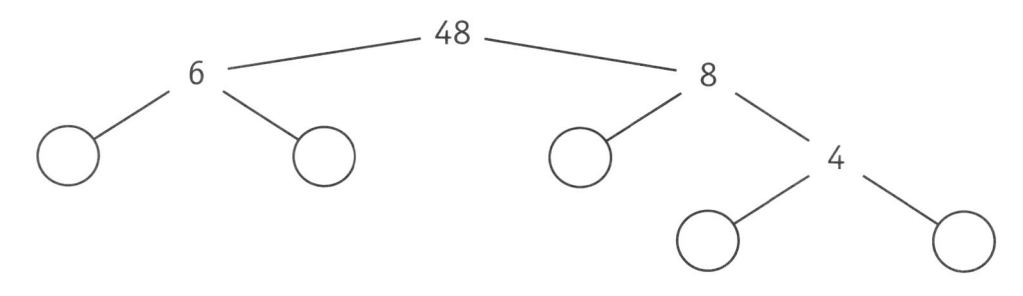

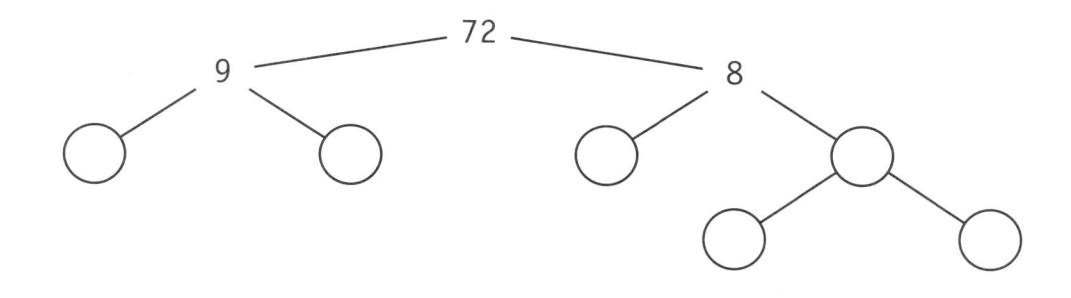

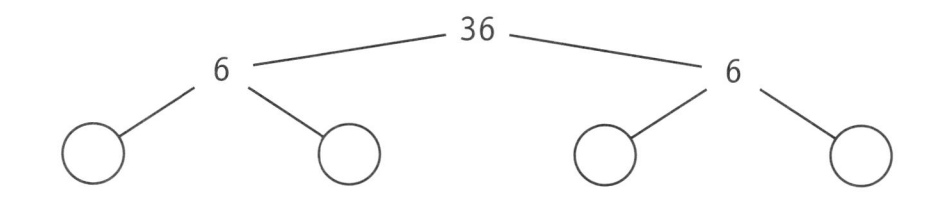

Activity 2: Write the first five *multiples* of each number. Start each list by rewriting the number itself.

EXAMPLE   6:   6   12   18   24   30

1. 4: ____ ____ ____ ____ ____
2. 10: ____ ____ ____ ____ ____
3. 7: ____ ____ ____ ____ ____
4. 5: ____ ____ ____ ____ ____
5. 3: ____ ____ ____ ____ ____

PART B
LEAST COMMON MULTIPLE PROBLEMS

Activity: We have set the two wheels on our cuckoo clock to turn at different rates. Your job is to figure out the least common multiple. It is the time when the wheels meet and the dancer comes out of the clock.

1. number of hours for one complete turn: 6 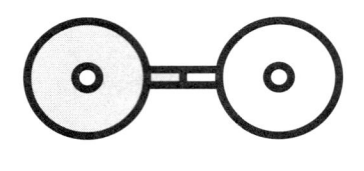
   number of hours for one complete turn: 9

   What is the least common multiple? ____

2. number of hours for one complete turn: 7 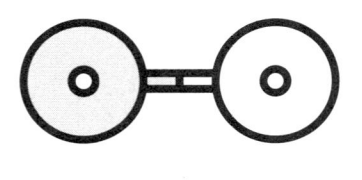
   number of hours for one complete turn: 4

   What is the least common multiple? ____

3. number of hours for one complete turn [5]   number of hours for one complete turn [8]

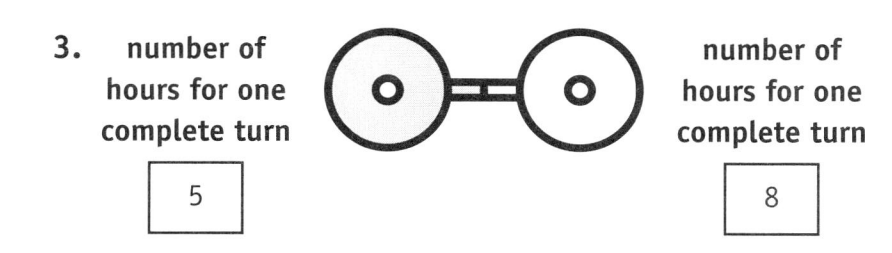

What is the least common multiple? _____

4. number of hours for one complete turn [3]   number of hours for one complete turn [10]

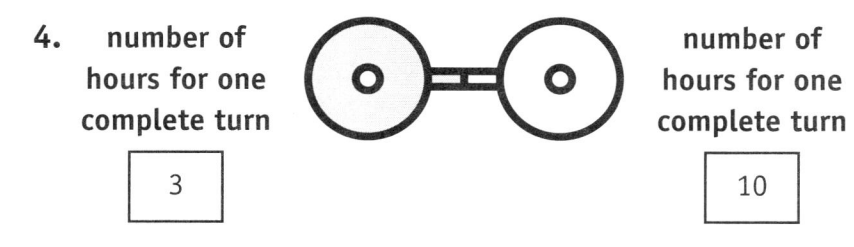

What is the least common multiple? _____

PART C
LINES OF SYMMETRY

Activity: Draw a line of symmetry through each of the following objects.

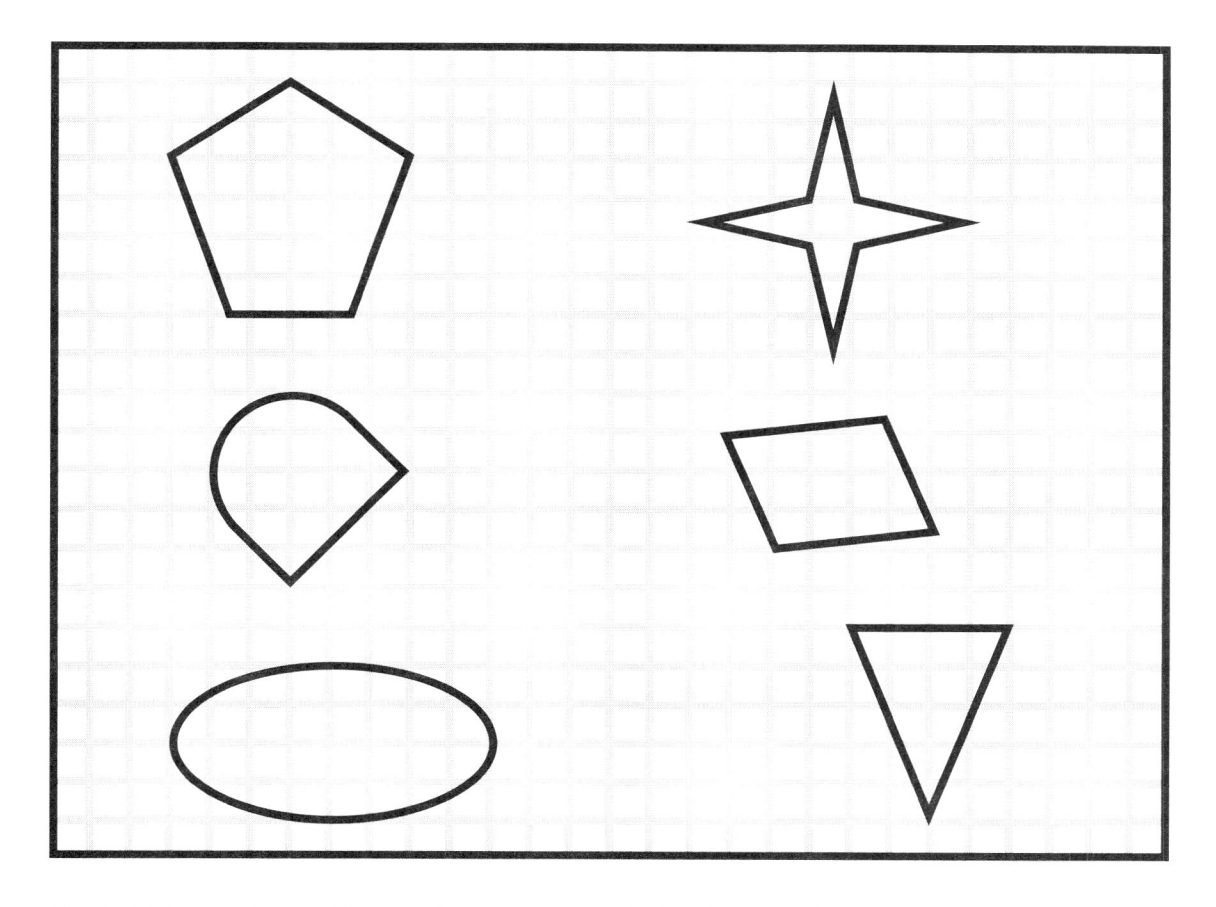

MOBILES—MOVING ART
Practice With LCM

LESSON 97

Name _____

PART A
WARM-UP ACTIVITY: EXPONENTS

Activity: Rewrite the following numbers using exponents.

1. $4 \times 4 \times 4 \times 4 \times 4 =$ _____

2. $3 \times 3 \times 3 \times 3 \times 5 =$ _____

3. $2 \times 2 \times 9 \times 9 \times 10 =$ _____

4. $5 \times 5 \times 4 \times 4 \times 3 \times 3 \times 2 \times 2 \times 1 \times 1 =$ _____

5. $1{,}450 \times 1{,}450 \times 1{,}450 \times 1{,}450 =$ _____

PART B
LEAST COMMON MULTIPLES

Activity: You worked on a problem like this in the last lesson. When will the wheels on the cuckoo clock meet and the dancer come out of the clock?

1. number of hours for one complete turn: 2

   number of hours for one complete turn: 8

   What is the least common multiple? _____

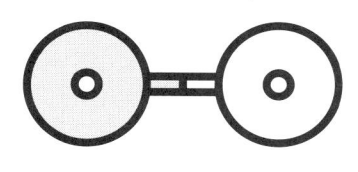

2. number of hours for one complete turn: 3

   number of hours for one complete turn: 7

   What is the least common multiple? _____

Student Workbook • Number Sense   Unit 5 Factors, Patterns, and Multiples   **279**

3. number of hours for one complete turn: 6

   number of hours for one complete turn: 10

   What is the least common multiple? _____

4. number of hours for one complete turn: 5

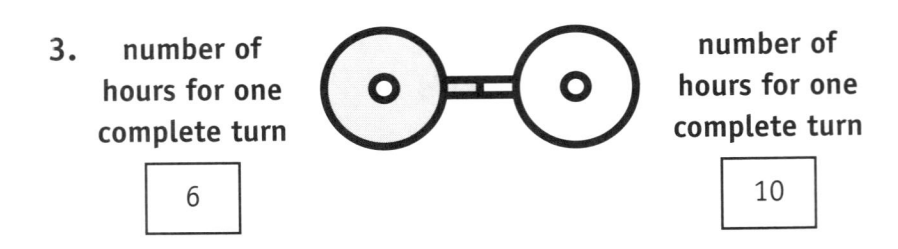

   number of hours for one complete turn: 15

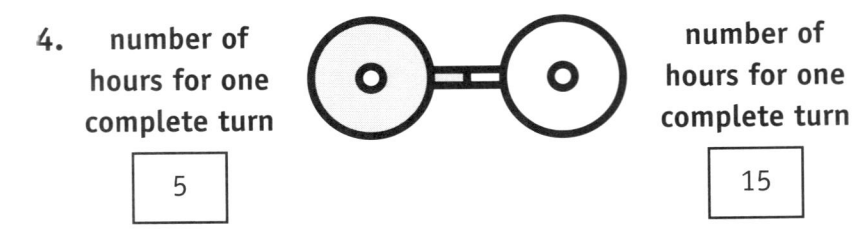

   What is the least common multiple? _____

PART C
DESIGNING MOBILES

Activity: Draw a sketch of a mobile on the grid. You don't have to be exact because you will be able to redraw your design on the next page. Use one or more types of objects for your mobile. Show slides, flips, and/or turns for the objects. Decide whether the mobile will or will not have symmetry.

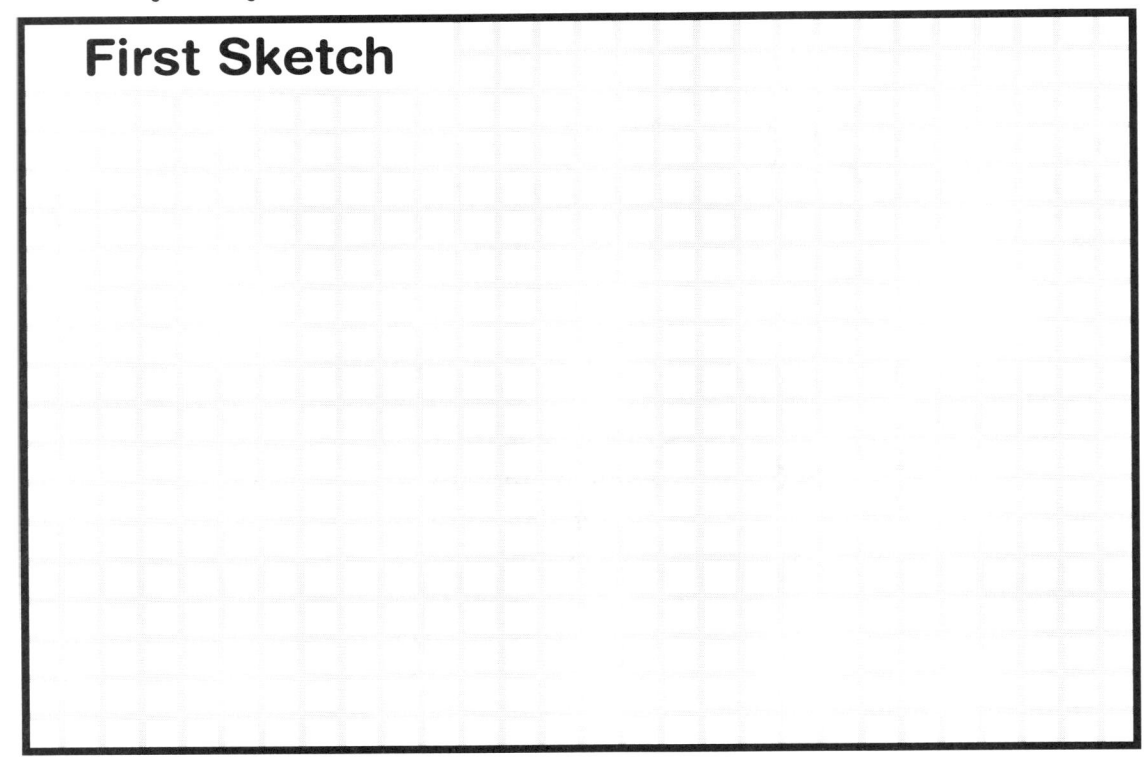

First Sketch

Final Drawing

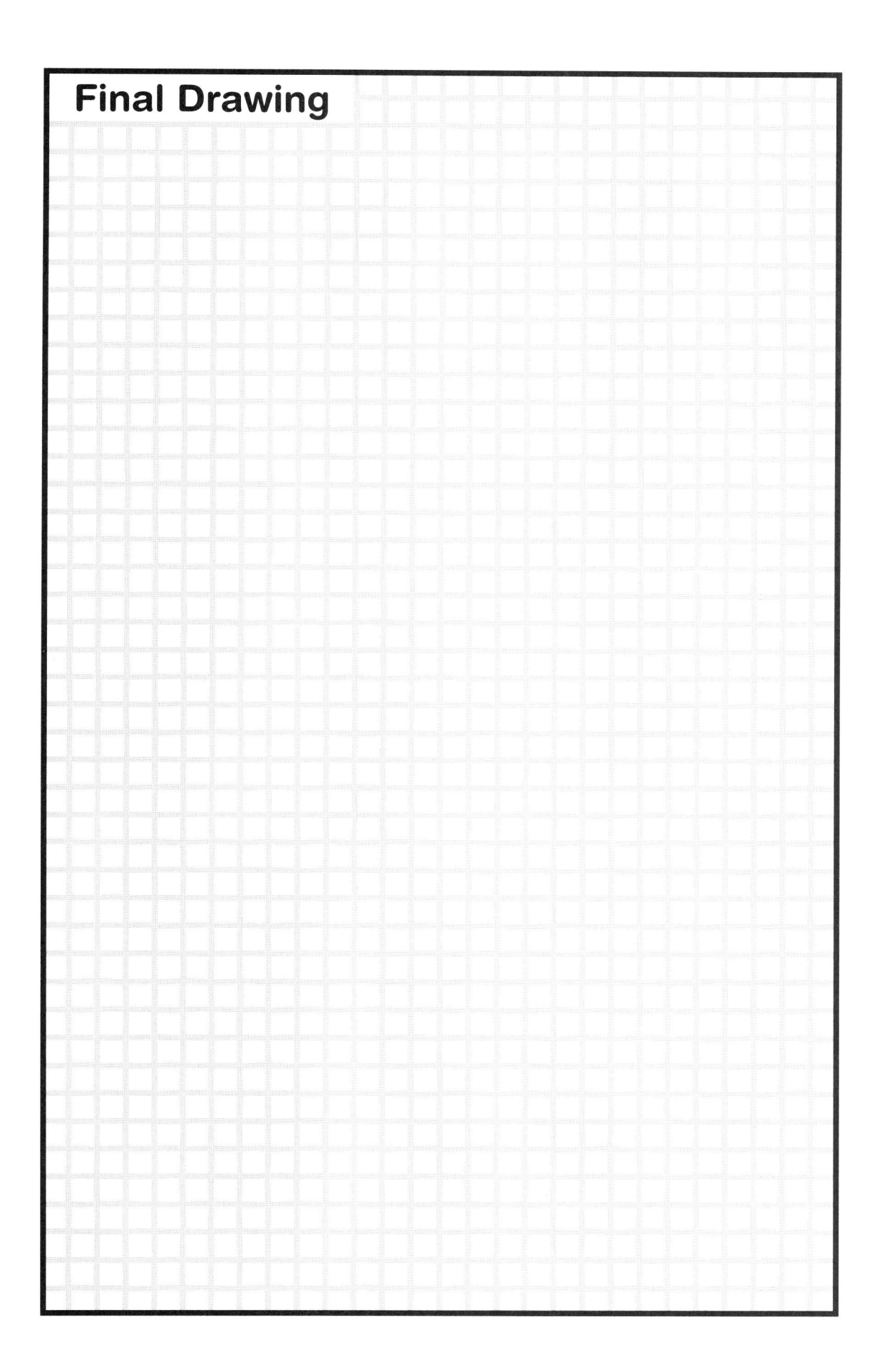

ADDING A THIRD WHEEL TO THE CUCKOO CLOCK
Rotational Symmetry

LESSON 98

Name _____

PART A
WARM-UP ACTIVITY: LOOKING FOR NUMBERS

Activity: Complete Problems 1-5 using this table of numbers from 1 to 100.

1	2	3	4	5	6	7	8	9	10
11	12	13	14	15	16	17	18	19	20
21	22	23	24	25	26	27	28	29	30
31	32	33	34	35	36	37	38	39	40
41	42	43	44	45	46	47	48	49	50
51	52	53	54	55	56	57	58	59	60
61	62	63	64	65	66	67	68	69	70
71	72	73	74	75	76	77	78	79	80
81	82	83	84	85	86	87	88	89	90
91	92	93	94	95	96	97	98	99	100

1. List seven prime numbers: _____

2. List three composite numbers: _____

3. List three square numbers: _____

4. List four numbers that are powers of 10: _____

5. List four numbers that you can divide by 6: _____

Student Workbook • *Number Sense*   Unit 5 Factors, Patterns, and Multiples

PART B
LEAST COMMON MULTIPLES

Activity: In this activity we have added a third wheel to our cuckoo clock. Now we have a gray wheel, a white wheel, and a black wheel. The clips for *all three wheels* have to touch for the Russian dancer to come out and dance. You need to think about all three numbers to find the least common multiple.

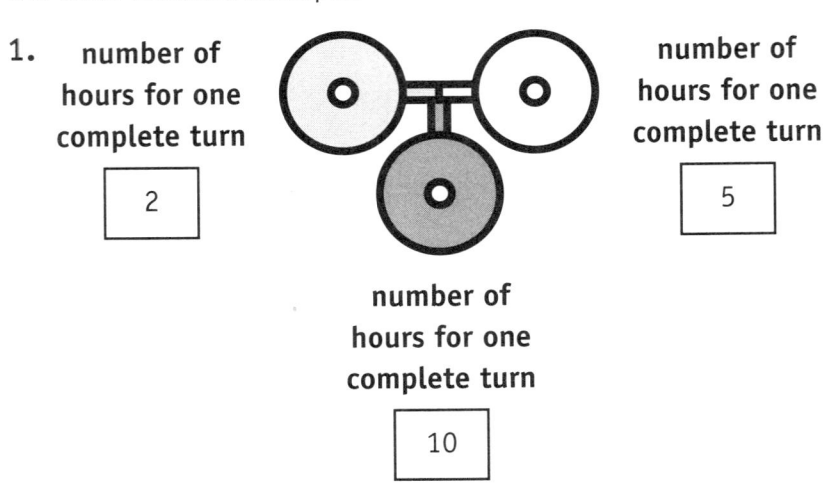

1. number of hours for one complete turn: 2

    number of hours for one complete turn: 5

    number of hours for one complete turn: 10

    What is the least common multiple? _____

2. number of hours for one complete turn: 4

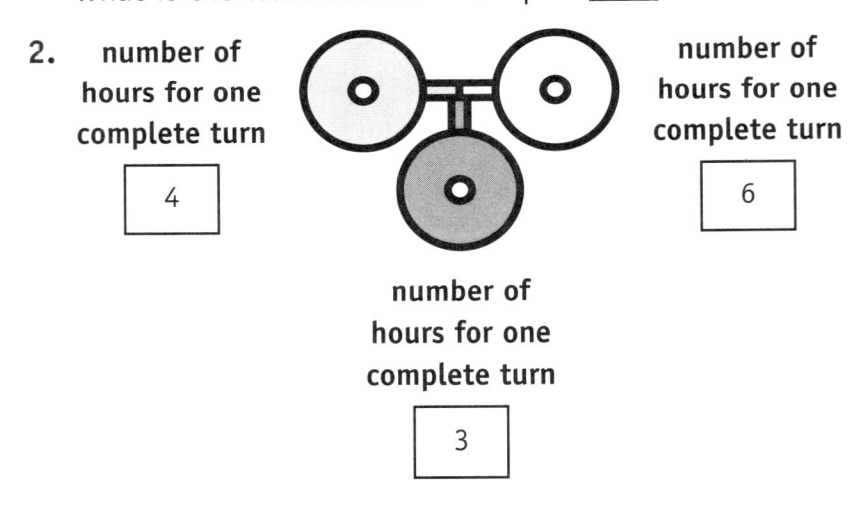

    number of hours for one complete turn: 6

    number of hours for one complete turn: 3

    What is the least common multiple? _____

3. number of hours for one complete turn

| 2 |

number of hours for one complete turn

| 2 |

number of hours for one complete turn

| 7 |

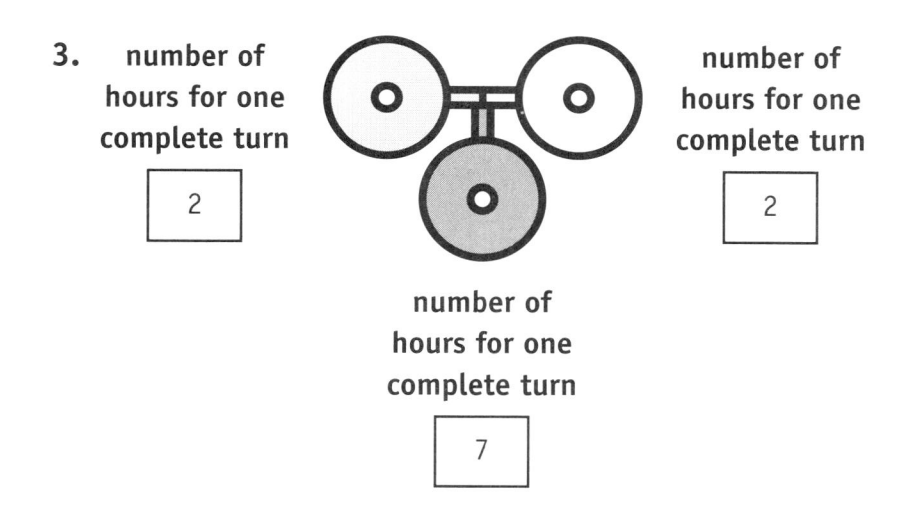

What is the least common multiple? _____

PART C
WHICH OBJECTS HAVE ROTATIONAL SYMMETRY?

Activity 1: Circle the objects that have rotational symmetry.

Activity 2: Select one of the objects from Activity 1 that *does not* have rotational symmetry. Draw it on the grid so that it has rotational symmetry.

NUMBER SENSE
Summing Up

LESSON 99

Name _____

PART A
WARM-UP ACTIVITY: YES OR NO

Activity: Circle YES or NO for each statement.

1. 17 is a prime number. — YES   NO
2. 28,796 is an even number. — YES   NO
3. $10 \times 10 \times 10 = 10^6$ — YES   NO
4. 22 is a prime number. — YES   NO
5. The factors of 10 are 1, 2, 5, and 10. — YES   NO
6. 8 is a square number. — YES   NO
7. $87 = 80 + 7$ — YES   NO
8. 27 is an odd number. — YES   NO
9. $80 = 8 \times 10$ — YES   NO
10. $4 \times 4 \times 4 \times 2 \times 2 = 4^3 \times 2^2$ — YES   NO

PART B
HOW MANY WAYS CAN YOU DESCRIBE A NUMBER?

Activity: For each of the following numbers, show at least four different ways of writing the number. Use what you've learned about odd, even, prime, and composite numbers, dividing rules, multiples, factors, and patterns in addition and multiplication. The example gives some sample answers.

EXAMPLE

14
$14 = 10 + 4$
It is an even number.
It is a composite number.
We can divide by 2.
It is a multiple of 7.
It is a multiple of 2.

36

9

37

90

PART C
CHOOSE A NUMBER

Activity: Choose a number from the grid that you think is interesting. Then answer the questions that follow the grid.

1	2	3	4	5	6	7	8	9	10
11	12	13	14	15	16	17	18	19	20
21	22	23	24	25	26	27	28	29	30
31	32	33	34	35	36	37	38	39	40
41	42	43	44	45	46	47	48	49	50
51	52	53	54	55	56	57	58	59	60
61	62	63	64	65	66	67	68	69	70
71	72	73	74	75	76	77	78	79	80
81	82	83	84	85	86	87	88	89	90
91	92	93	94	95	96	97	98	99	100

What is the number? _____

What do you find interesting about this number? _____

_____

_____

_____

_____

UNIT 5 REVIEW

Name _____

LESSON 100

PART A
ADDITION PATTERNS FOR EVEN AND ODD NUMBERS

Activity: Without solving, tell whether the answer to each of the following addition problems is going to be *odd* or *even*. Circle the correct response. Check with a calculator.

1. 14 + 32 = ____   ODD   or   EVEN
2. 429 + 577 = ____   ODD   or   EVEN
3. 868 + 399 = ____   ODD   or   EVEN
4. 6,987 + 2,888 = ____   ODD   or   EVEN

PART B
MULTIPLICATION PATTERNS FOR EVEN AND ODD NUMBERS

Activity: Without solving, tell whether the answer to each of the following multiplication problems is going to be *odd* or *even*. Circle the correct response. Check with a calculator.

1. 34 * 52 = ____   ODD   or   EVEN
2. 429 * 177 = ____   ODD   or   EVEN
3. 368 * 299 = ____   ODD   or   EVEN
4. 3,987 * 4,888 = ____   ODD   or   EVEN

PART C
IDENTIFYING SQUARE NUMBERS

Activity: There is one square number in each of these problems. Circle the numbers that are square.

1. 15, 25, 35, 45
2. 16, 26, 32, 46
3. 19, 29, 39, 49
4. 81, 91, 101, 111

Student Workbook • Number Sense   Unit 5 Factors, Patterns, and Multiples

PART D
TRIANGULAR NUMBERS

Activity: Use squares to draw these triangular numbers.

EXAMPLE   3

Answer: ⌐⌐

6	21	28

PART E
EXPONENTS

Activity: Rewrite the following using exponents.

EXAMPLE $10 \times 10 \times 10 =$ _____

Answer: 10^3

1. $7 \times 7 \times 7 =$ _____

2. $3 \times 3 =$ _____

3. $5 \times 5 \times 5 \times 5 =$ _____

4. $10 \times 10 \times 10 \times 10 \times 10 \times 10 \times 10 \times 10 \times 10 \times 10 \times 10 =$ _____

5. $2 \times 2 \times 2 =$ _____

PART F
COMMON MULTIPLES

Activity: From the lists of multiples for each problem, list some common multiples for the numbers given.

1. What are some common multiples of 5 and 15? _____

Multiples of 5	5	10	15	20	25	30	35	40	45
Multiples of 15	15	30	45						

2. What are some common multiples of 4 and 6? _____

Multiples of 4	4	8	12	16	20	24	28	32	36
Multiples of 6	6	12	18	24	30	36			

PART G
FLIPS, SLIDES, AND TURNS

Activity: For each drawing, tell whether it's a picture of a slide, a flip, or a turn.

1.

This is a _____.

2.

This is a _____.

PART H
LEAST COMMON MULTIPLES

Activity: Find the least common multiple (LCM) for the numbers in each problem.

1. What is the least common multiple of 3 and 6?

3	3	6	9	12	15	18	21	24	27	30
6	6	12	18	24	30	36	42	48	60	

2. What is the least common multiple of 2, 4, and 6?

2	2	4	6	8	10	12	14	16	18	20	22	24	26	28	30
4	4	8	12	16	20	24	28	32	36	40					
6	6	12	18	24	30	36	42	48	60						

PART I
SYMMETRY

Activity: Tell whether the following objects have symmetry. Circle YES or NO.

1. This object has rotational symmetry.　　　　YES　　NO

2. This object has rotational symmetry.　　　　YES　　NO

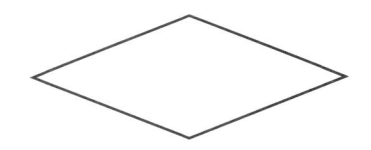

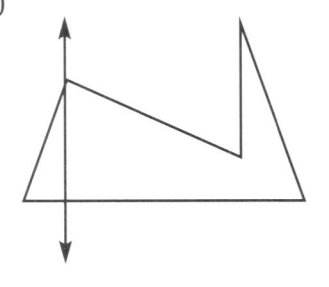

3. This object has a line of symmetry.　　　　YES　　NO

4. This object has a line of symmetry.   YES   NO

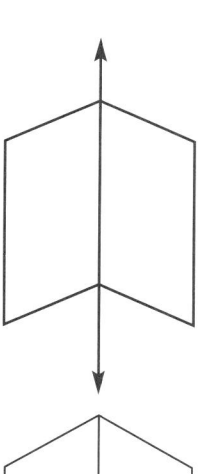

5. This object has rotational symmetry.   YES   NO